CHANGJIAN MUCAO YUANSE TUPU

常见牧草原色图谱

编 著 者

李峻成
高崇岳
(澳)李光棣

金盾出版社

内 容 提 要

全书按植物分类学惯例排序，收编常见牧草植物432种，分属67科，同时收集了常见的有毒有害植物19种，书后附录中文索引。编著者本着实用的宗旨，采用图文对照形式编排。牧草有单株或群体、花序或花、果或果穗、根、茎、叶的照片，力求显示出种的识别特征，并附有形态描述、繁殖方式及家畜采食特性与牧草评价。增强了读者感性认知的效果，查阅方便，适合从事草业科学的同行，特别是初涉草业界的新人，查阅参考。

图书在版编目（CIP）数据

常见牧草原色图谱/李峻成、高崇岳，（澳）李光棣编著. --北京：金盾出版社，2010.5

ISBN 978-7-5082-6259-8

Ⅰ. ①常… Ⅱ. ①李…②高…③李… Ⅲ. ①牧草-中国-图谱 Ⅳ. ①S54-64

中国版本图书馆 CIP 数据核字（2010）第 039436 号

金盾出版社出版、总发行
北京太平路5号（地铁万寿路站往南）
邮政编码：100036　电话：68214039　83219215
传真：68276683　网址：www.jdcbs.cn
北京蓝迪彩色印务有限公司印刷
北京蓝迪彩色印务有限公司装订
各地新华书店经销
开本：787×1092　1/16　印张：15.625　字数：300千字
2010年5月第1版第1次印刷
印数：1~8 000册　定价：59.00元

（凡购买金盾出版社的图书，如有缺页、
　倒页、脱页者，本社发行部负责调换）

前　言

绿色植物是生物"营养链"的基础,是动物生命的营养源。粮食作物直接提供了人类能量的大部分;而非粮作物——牧草给人类的食物提供了更高的营养物质。牧草是一个生物能源的加工厂;牧草是一条蛋白质生产线;牧草是防止水土流失的天然屏障;牧草同样是地球环境的卫士,为人类的生活提供了五彩缤纷的世界。总之,牧草对于人类的重要性丝毫不逊色于粮食,牧草通过家畜"有机加工"变成了人类高级的肉、蛋、奶食品,西方国家很早就有"草就是肉"的说法。作者在长期从事基层草地农业研究、教学、推广和生产实践中,认识到牧草植物资源的科学利用和保护对于草业科学的发展是十分重要的。而利用和保护的必要前提是认知和识别它们,认知和识别又是一门实践性很强的工作。

《常见牧草原色图谱》共收集我国常见栽培与野生牧草植物 432 种,分属 67 科。以豆科、禾本科、菊科等为主,还收集了 19 种常见的毒害植物,以备生产中识别。本书以图文并茂的形式,采用牧草全株或主要器官的原色图谱,简述其植物学特征、生长环境、分布地域、家畜利用方式和饲用价值等,还配有简短的英文介绍。目的是为牧草、家畜、植物等学科,以及从事生态建设、农牧业生产的教学与研究人员提供帮助。

由于我们经验不足,学识有限,加之涉及地域广泛、拍照季节性强等多方面困扰,书中错误在所难免,恳请同行和读者给予批评指正!

在编写过程中,兰州大学孙继周老师、刘照辉老师、常生华老师、甘肃省草原工作总站赵怀德老师、兰州商学院高维宁和庆阳黄土高原试验站冯治山、李太湖、姜亮东等同志给予了热情的帮助,特别是聂斌同志提供了 20 余张青藏高原的照片,对他们所付出的辛勤劳动表示衷心感谢!

本书由兰州大学草地农业科技学院李峻成主编,兰州大学草地农业科技学院高崇岳撰写,GuangDilI, E.H.Graham Centre for Agricultural Inrovation (Alliance between NSW Departrnent of Primary Industries and Charles Sturt University Australia)英校对。希望广大读者提出宝贵意见。

编著者
2009. 7

目　录

一　木贼科
1　问荆 …………………………………… 1
2　散生木贼 ……………………………… 1
3　节节草 ………………………………… 2

二　满江红科
4　细绿萍 ………………………………… 2

三　麻黄科
5　草麻黄 ………………………………… 3

四　鸭跖草科
6　鸭跖草 ………………………………… 3

五　桑科
7　葎草 …………………………………… 4
8　大麻 …………………………………… 4

六　荨麻科
9　大蝎子草 ……………………………… 5

七　蓼科
10　旱型两栖蓼 …………………………… 5
11　东方蓼 ………………………………… 6
12　酸模叶蓼 ……………………………… 6
13　绵毛酸模叶蓼 ………………………… 7
14　萹蓄 …………………………………… 7
15　白花蓼 ………………………………… 8
16　西伯利亚蓼 …………………………… 8
17　尼泊尔蓼 ……………………………… 9
18　珠芽蓼 ………………………………… 9
19　柳叶刺蓼 ……………………………… 10
20　黏毛蓼 ………………………………… 10
21　杠板归 ………………………………… 11
22　羊蹄 …………………………………… 11
23　鲁梅克斯 ……………………………… 12

八　藜科
24　灰绿碱蓬 ……………………………… 12
25　绳虫实 ………………………………… 13
26　合头草 ………………………………… 13
27　雾冰藜 ………………………………… 14
28　菊叶香藜 ……………………………… 14
29　盐生草 ………………………………… 15
30　白茎盐生草 …………………………… 15

31　刺藜 …………………………………… 16
32　尖头叶藜 ……………………………… 16
33　灰绿藜 ………………………………… 17
34　地肤 …………………………………… 17
35　碱地肤 ………………………………… 18
36　沙蓬 …………………………………… 18
37　大苞滨藜 ……………………………… 19
38　中亚滨藜 ……………………………… 19
39　鞑靼滨藜 ……………………………… 20
40　盐爪爪 ………………………………… 20
41　刺沙蓬 ………………………………… 21
42　猪毛菜 ………………………………… 21
43　兴安虫实 ……………………………… 22
44　角果碱蓬 ……………………………… 22
45　小藜 …………………………………… 23
46　藜 ……………………………………… 23
47　杂配藜 ………………………………… 24
48　梭梭柴 ………………………………… 24

九　苋科
49　牛膝 …………………………………… 25
50　籽粒苋 ………………………………… 25
51　繁穗苋 ………………………………… 26
52　反枝苋 ………………………………… 26
53　皱果苋 ………………………………… 27
54　美国籽粒苋 …………………………… 27

十　马齿苋科
55　马齿苋 ………………………………… 28

十一　石竹科
56　石竹 …………………………………… 28
57　牛繁缕 ………………………………… 29
58　瞿麦 …………………………………… 29
59　麦瓶草 ………………………………… 30
60　女娄菜 ………………………………… 30
61　蝇子草 ………………………………… 31

十二　毛茛科
62　展枝唐松草 …………………………… 31
63　瓣蕊唐松草 …………………………… 32
64　东方铁线莲 …………………………… 32
65　甘青铁线莲 …………………………… 33

66	大火草	33
67	升麻	34
68	绵团铁线莲	34

十三 罂粟科

69	角茴香	35
70	野罂粟	35
71	多刺绿绒蒿	36

十四 十字花科

72	荠菜	36
73	离子芥	37
74	弯曲碎米荠	37
75	播娘蒿	38
76	小果亚麻芥	38
77	遏蓝菜	39
78	独行菜	39
79	宽叶独行菜	40
80	风花菜	40
81	串珠芥	41
82	垂果蒜芥	41
83	桂竹糖芥	42
84	垂果南芥	42

十五 景天科

85	费菜	43

十六 鼠李科

86	酸枣	43

十七 蔷薇科

87	蛇莓	44
88	委陵菜	44
89	匍枝委陵菜	45
90	朝天委陵菜	45
91	二裂委陵菜	46
92	多茎委陵菜	46
93	鹅绒委陵菜	47
94	西山委陵菜	47
95	龙牙草	48
96	水杨梅	48
97	地榆	49
98	绒毛绣线菊	49
99	珍珠梅	50
100	悬钩子	50
101	扁核木	51
102	鲜卑花	51
103	金露梅	52
104	银露梅	52

十八 豆科

105	陇东苜蓿	53
106	紫花苜蓿	53
107	白花苜蓿	54
108	黄花苜蓿	54
109	圆形苜蓿	55
110	蒺藜状苜蓿	55
111	红豆草	56
112	红三叶	56
113	白三叶	57
114	沙打旺	57
115	黄花草木樨	58
116	白花草木樨	58
117	多变小冠花	59
118	百脉根	59
119	紫云英	60
120	箭筈豌豆	60
121	毛苕子	61
122	美国香豌豆	61
123	山羊豆	62
124	山黧豆	62
125	花棒	63
126	紫穗槐	63
127	豌豆	64
128	扁豆	64
129	蚕豆	65
130	莲山黄芪	65
131	鸡峰山黄芪	66
132	糙叶黄芪	66
133	草木樨状黄芪	67
134	太白岩黄芪	67
135	野大豆	68
136	野绿豆	68
137	歪头菜	69
138	甘草	69
139	米口袋	70
140	狭叶米口袋	70
141	少花米口袋	71
142	长叶铁扫帚	71
143	截叶铁扫帚	72
144	达乌里胡枝子	72
145	天蓝苜蓿	73
146	扁蓿豆	73

147	花苜蓿	74
148	披针叶黄花	74
149	洛氏锦鸡儿	75
150	白皮锦鸡儿	75
151	中间锦鸡儿	76
152	小叶锦鸡儿	76
153	二色棘豆	77
154	甘肃棘豆	77
155	苦豆子	78
156	苦马豆	78
157	三齿萼野豌豆	79
158	山野豌豆	79
159	小巢菜	80
160	木蓝	80
161	合欢	81
162	白花刺	81
163	红花岩黄芪	82
164	美丽胡枝子	82
165	蓝花棘豆	83
166	青海黄芪	83
167	毛胡枝子	84

十九　酢浆草科

168	酢浆草	84
169	铜锤草	85
170	黄花酢浆草	85

二十　牻牛儿苗科

171	牻牛儿苗	86
172	鼠掌老鹳草	86

二十一　蒺藜科

173	蒺藜	87
174	蝎虎霸王	87
175	白刺	88
176	唐古特白刺	88
177	葡根骆驼蓬	89
178	骆驼蓬	89

二十二　远志科

179	细叶远志	90
180	西伯利亚远志	90

二十三　大戟科

181	铁苋菜	91
182	地锦	91
183	地构叶	92

二十四　柽柳科

184	红柳	92
185	枇杷柴	93

二十五　胡颓子科

186	沙棘	93
187	沙枣	94

二十六　五加科

188	五加	94

二十七　葫芦科

189	栝楼	95

二十八　锦葵科

190	苘麻	95
191	野西瓜苗	96
192	圆叶锦葵	96
193	蜀葵	97
194	野葵	97

二十九　藤黄科

195	长柱金丝桃	98

三十　堇菜科

196	早开堇菜	98
197	紫花地丁	99
198	犁头草	99

三十一　柳叶菜科

199	柳叶菜	100
200	月见草	100

三十二　伞形科

201	野胡萝卜	101
202	硬阿魏	101
203	水芹	102
204	柴胡	102
205	窃衣	103
206	藁本	103
207	石防风	104

三十三　报春花科

208	狼尾珍珠菜	104
209	羽叶点地梅	105

三十四　白花丹科

210	金色补血草	105
211	二色补血草	106

三十五　龙胆科

212	龙胆	106
213	獐牙菜	107
214	秦艽	107

三十六	夹竹桃科
215	大叶罗布麻 … 108

三十七	萝藦科
216	鹅绒藤 … 108
217	地梢瓜 … 109
218	杠柳 … 109

三十八	旋花科
219	打碗花 … 110
220	田旋花 … 110
221	银灰旋花 … 111
222	藤长苗 … 111
223	圆叶牵牛 … 112
224	中国菟丝子 … 112
225	日本菟丝子 … 113

三十九	紫草科
226	聚合草 … 113
227	附地菜 … 114
228	鹤虱 … 114
229	狼紫草 … 115
230	倒提壶 … 115
231	微孔草 … 116

四十	马鞭草科
232	马鞭草 … 116

四十一	唇形科
233	香薷 … 117
234	密花香薷 … 117
235	水棘针 … 118
236	夏至草 … 118
237	益母草 … 119
238	细叶益母草 … 119
239	紫苏 … 120
240	百里香 … 120
241	冬青兔唇花 … 121
242	甘西鼠尾草 … 121
243	鼬瓣花 … 122
244	糙苏 … 122
245	香青兰 … 123
246	丹参 … 123
247	薄荷 … 124
248	并头黄芩 … 124
249	黄芩 … 125

四十二	茄科
250	枸杞 … 125
251	黑果枸杞 … 126
252	挂金灯酸浆 … 126
253	龙葵 … 127
254	红果龙葵 … 127
255	青杞 … 128

四十三	玄参科
256	柳穿鱼 … 128
257	匍茎通泉草 … 129
258	通泉草 … 129
259	北水苦荬 … 130
260	甘肃马先蒿 … 130
261	轮叶马先蒿 … 131
262	半扭卷马先蒿 … 131
263	藓状马先蒿 … 132
264	小米草 … 132
265	地黄 … 133
266	阿拉伯婆婆纳 … 133
267	光药大黄花 … 134

四十四	紫葳科
268	红花角蒿 … 134
269	黄花角蒿 … 135

四十五	列当科
270	列当 … 135

四十六	车前科
271	车前 … 136
272	大车前 … 136
273	小车前 … 137
274	平车前 … 137

四十七	茜草科
275	茜草 … 138
276	猪殃殃 … 138
277	蓬子菜 … 139

四十八	败酱科
278	异叶败酱 … 139
279	缬草 … 140

四十九	桔梗科
280	桔梗 … 140
281	石沙参 … 141
282	柳叶沙参 … 141
283	秦岭沙参 … 142

五十	菊科
284	串叶松香草 … 142

285	野茼蒿	143	329	黄缨菊 … 165
286	鬼针草	143	330	聚头蓟 … 165
287	小花鬼针草	144	331	黄帚橐吾 … 166
288	腺梗豨莶	144	332	箭叶橐吾 … 166
289	飞廉	145	333	狗舌草 … 167
290	天名精	145	334	蒙古雅葱 … 167
291	刺儿菜	146	335	艾蒿 … 168
292	大蓟	146	336	黄花蒿 … 168
293	菊苣	147	337	黄蒿 … 169
294	小飞蓬	147	338	细裂叶莲蒿 … 169
295	旱莲草	148	339	大籽蒿 … 170
296	一年蓬	148	340	黑沙蒿 … 170
297	辣子草	149	341	蒙古蒿 … 171
298	泥胡菜	149	342	茭蒿 … 171
299	蓼子朴	150	343	冷蒿 … 172
300	山苦荬	150	344	耆状亚菊 … 172

五十一 香蒲科
345 宽叶香蒲 … 173

301	抱茎苦荬菜	151
302	蒙山莴苣	151
303	兴安毛莲菜	152
304	阿尔泰狗娃花	152
305	裂叶马兰	153
306	花花柴	153
307	风毛菊	154
308	鳍蓟菊	154
309	顶羽菊	155
310	野菊	155
311	蒲公英	156
312	苦荬菜	156
313	苍耳	157
314	款冬	157
315	黄鹌菜	158
316	大丁草	158
317	火绒草	159
318	秋鼠麹草	159
319	伞花雅葱	160
320	拐轴雅葱	160
321	旋覆花	161
322	麻花头	161
323	牛蒡	162
324	菊芋	162
325	漏芦	163
326	丝裂亚菊	163
327	细叶亚菊	164
328	大花千里光	164

五十二 禾本科
346	老芒麦	173
347	垂穗披碱草	174
348	无芒雀麦	174
349	多节雀麦	175
350	冰草	175
351	䕡草	176
352	硬质早熟禾	176
353	高羊茅	177
354	紫羊茅	177
355	猫尾草	178
356	鸭茅	178
357	燕麦	179
358	野燕麦	179
359	羊茅	180
360	多年生黑麦草	180
361	黑麦	181
362	早熟禾(瓦巴斯)	181
363	一年生早熟禾	182
364	高山早熟禾	182
365	臭草	183
366	赖草	183
367	白羊草	184
368	虎尾草	184
369	中间偃麦草	185

370	高冰草	185
371	茇茇草	186
372	柳枝稷	186
373	野稷	187
374	湖南稷子	187
375	升马唐	188
376	止血马唐	188
377	狗牙根	189
378	牛筋草	189
379	白茅	190
380	无芒稗	190
381	长芒野稗	191
382	西来稗	191
383	䅟草	192
384	大画眉草	192
385	小画眉草	193
386	苇状看麦娘	193
387	狼尾草	194
388	金色狗尾草	194
389	狗尾草	195
390	大狗尾草	195
391	谷莠子	196
392	荻	196
393	芦苇	197
394	假苇拂子茅	197
395	长芒草	198
396	黄背草	198
397	苏丹草	199
398	先锋高丹草	199
399	甜高粱	200

五十三 莎草科

400	白颖苔草	200
401	香附子	201
402	蔗草	201
403	水莎草	202
404	旋鳞莎草	202
405	两歧飘拂草	203

五十四 天南星科

406	石菖蒲	203

五十五 忍冬科

407	羽裂楚子蔗	204
408	金银花	204

五十六 百合科

409	大苞萱草	205
410	卷丹	205
411	天蒜	206
412	碱韭	206
413	秦岭野韭	207
414	野韭	207
415	天门冬	208
416	戈壁天门冬	208
417	芦笋	209
418	百合	209
419	山丹花	210

五十七 鸢尾科

420	马蔺	210
421	射干鸢尾	211
422	膜苞鸢尾	211

五十八 兰科

423	绶草	212

五十九 亚麻科

424	宿根亚麻	212

六十 荷包牡丹科

425	地丁草	213

六十一 虎耳草科

426	梅花草	213

六十二 川续断科

427	东北川续断	214

六十三 薯蓣科

428	穿龙薯蓣	214

六十四 檀香科

429	百蕊草	215

六十五 商陆科

430	美国商陆	215

六十六 无患子科

431	文冠果	216

六十七 木犀科

432	迎春花	216

附录一：有毒有害植物

一 豆科

1	黄毛棘豆	217
2	苦参	217

二 毛茛科

3	耧斗菜	218
4	茴茴蒜	218

5　草玉梅 ………………… 219
　　6　腺毛唐松草 …………… 219
　　7　翠雀 …………………… 220
　　8　腺毛翠雀 ……………… 220
　　9　石龙芮 ………………… 221
三　罂粟科
　　10　秃疮花 ………………… 221
　　11　灰绿黄堇 ……………… 222
　　12　白屈菜 ………………… 222
四　大戟科
　　13　泽漆 …………………… 223
　　14　甘遂 …………………… 223
　　15　大戟 …………………… 224
五　瑞香科
　　16　狼毒 …………………… 224
六　茄科
　　17　曼陀罗 ………………… 225
七　天南星科
　　18　半夏 …………………… 225
八　禾本科
　　19　醉马草 ………………… 226

附录二：中文名索引 …………………… 227
参考文献 ………………………………… 237

一　木贼科

1　问　荆

学名　*Equisetum arvense* L.
英名　Common Horsetail

多年生草本，依靠根茎和孢子繁殖。地下根茎发达，并常具小球茎。地上茎直立，两型。孢子茎先发，不分枝，高 10～30 厘米，肉质，黄白色或淡黄褐色，鞘长而大，棕褐色；孢子囊穗顶生，钝头。孢子茎枯萎前，在同一根茎上生出营养茎，茎高 50 厘米左右，绿色，有棱脊 10～15 条；鞘齿披针形，黑色，边缘膜质，灰白色；分枝轮生，单一或再分枝。属于优良牧草，绵羊最为喜食，马、牛、山羊、猪均采食。

A perennial herbaceous plant, propagating by rhizome and by spore. It occurs in rather moist farmlands and waste lands, and is distributed mainly in north and northeast China as well as Gansu and Shaanxi. The feeding value is moderate. Cattle, sheep and goats will graze it.

1-2　问荆植株　　　1-1　问荆

2　散生木贼

学名　*Equisetum diffusum* Don
英名　Common Scouring Rush

别名笔头草。多年生草本，依靠根茎或孢子繁殖。茎一型，直立或斜升，高 20～40 厘米，黄绿色，中心孔小或实，有棱脊 8～10 条，沿棱脊有小疣状突起；鞘片两侧隆起的棱角直达鞘齿的顶端，鞘齿长大而不脱落，先端和边缘深紫褐色；分枝轮生，小枝细而密。孢子囊穗着生于枝顶，圆柱形，钝头，孢子叶排列松散。常生于河边沙地或玉米、洋芋等旱作农田。早春和晚秋对牛、羊适口性较好，夏季草质坚韧，家畜不喜食。马单一采食过多，会引起尿结石。

A perennial herbaceous plant reproduced by rhizome and by spore. It occurs in sandy land, river sides and some farmlands. It is distributed over Loess plateau. The feeding value is moderate. Cattle, sheep and goats will graze it.

2-1　散生木贼　　　2-2　散生木贼孢子囊穗

3 节节草

学名 Hippochaete ramosissimum Desf.
英名 Branched Horgetail

多年生草本。茎直立或斜升,根茎黑褐色。基部分枝,分枝中空,绿色或灰绿色,茎有棱多条,粗糙。叶鞘黑色,三角形,基部具狭膜质白边。每节有 2~5 轮生分枝,有时上部不生小枝或仅生 1 小枝。孢子囊穗生于枝顶,长圆形,常有黑色小尖头。喜生于沙质土壤的农田、路边、水边,多种家畜喜食。

A perennial grass reproduced by rhizome and by spore. It can be found in sandy farmlands, creek banks, and wet lands of water sides. It is distributed throughout the country, particularly in northern part of China. The feeding value is moderate and cattle and sheep like to graze when it is fresh or dry.

3-1 节节草

3-2 节节草植株

二 满江红科

4 细绿萍

学名 Azolla imbricate (Roxb.)Nakai
英名 Mosquito Fern

细绿萍是红萍的一种,蕨类门植物,植株体细小,呈三角形,浮生于水面。根细长,根毛多数,悬垂于水平面下。生出新根叶时,老根脱落。茎细短,常有 6~8 个分枝。叶呈芝麻状,无叶柄,互生于各个分枝上,成覆瓦状。每叶分上下 2 片,上叶浮出水面,有叶绿体,可进行光合作用,下片叶沉入水中,呈鳞片状薄膜,红色或白色,能吸取营养和水分,子果着生于萍体的背面,初生为绿色,后呈红色或橙黄色。

4 细绿萍

A natant small herb, spreading by rhizome and by spore. It occurs in rice fields and ponds, distributed throughout the country with more popular in the provinces and autonomous regions south to the Yangze River.

三　麻黄科

5　草麻黄

学名　*Ephedra sinica* Stapf
英名　Chinese Ephedra

草本状小灌木。木质茎常横卧于地面，小枝丛生，直立，少分枝，高 20～40 厘米。叶基部结合成鞘，上部 2 裂，

5-4　草麻黄雄花与叶鞘　　5-3　草麻黄雄株　　5-2　草麻黄雌株叶鞘　　5-1　草麻黄雌株与雌球穗

裂片三角形。花单性，雌雄异株，雌球穗单一或成束，顶生或生于节上，卵形，苞片 4 对，绿色。果实成熟时，苞片膨大成肉质而变为红色；雄花单一，有梗或无梗，轮生或顶生，卵形或宽卵形。生于干旱草原和荒漠的石质低地山坡、沙滩、沙丘；分布在华北、西北及内蒙古。低等牧草，羊、骆驼乐食。

It is a shrubby herbaceous plant, grows on hilly, sloping fields and cliff edges. It is distributed over all of northern China, including the Loess Plateau. The feeding value is quite high. Sheep and camels will consume this plant as hay in winter.

四　鸭跖草科

6　鸭跖草

学名　*Commelina communis* L.
英名　Common Dayflower

一年生草本。茎下部匍匐生根，上部直立或斜生，有分枝。叶互生，披针形或卵状披针形，基部下延成鞘，有紫红色条纹。总苞片佛焰苞状，腋生，有长柄，宽心形，稍弯曲，边缘有硬毛。花数朵，略伸出苞外。3 花瓣中 2 片较大，深蓝色，较小的一片色淡。雄蕊 6，一半退化，先端成蝴蝶状。蒴果椭圆形，2 室，有种子 4 粒，种子表面粗糙，褐色。生长在农田、果园及周边荒地，家畜可食。

An annual loose-head herb reproduced from seeds. It grows in wet land, farmlands and grasslands. It is distributed from north to south in various regions east to Yunnan and Gansu provinces. This is good forages for cattle and pigs.

6-2　鸭趾草花序

6-1　鸭跖草

五 桑 科

7 葎草

学名 *Humulus scandens* L.
英名 Japanese Hop

7-1 葎草

7-2 葎草叶

7-3 葎草花序

别名降龙草、拉拉藤。一年生或多年生缠绕草本。茎枝和叶柄均有刚毛和倒钩刺。叶对生,叶片掌状5~7裂,叶面粗糙,疏生刚毛和白色腺点,背面沿叶脉及边缘具刚毛,有稀疏黄色小腺点,边缘具锯齿。花单性,雌雄异株,雄花序圆锥状,花小,雄花被片和雄蕊各5,黄绿色,雌花序穗状,腋生,花灰白色,花柱红褐色,通常十余朵花密集而下垂。瘦果扁圆形,淡黄色。因叶片具刚毛和刺,适口性较差,嫩枝叶切碎蒸煮后可饲喂猪和家禽。

A perennial or annual herb, commonly found nearby villages, roadsides and hills. It is distributed in south of the China and Shaanxi, Gansu, etc. The feeding value is low and pigs like to graze its tender branches and roots after boiled.

8 大 麻

学名 *Cannabis sativa* L.
英名 Hemp

8-1 大麻

8-2 大麻雄株   8-3 大麻雌株

一年生草本。高50~200厘米,茎秆直立粗壮,有纵棱,多分枝,具短腺毛,基部易木质化。掌状复叶,小叶披针形,边缘有锯齿。花单性,雌雄异株;雄花序圆锥状,雌花序球状或短穗状。瘦果卵状,有棱,种子深绿色。雄株早熟,韧皮纤维产量高。多种植在农田地边或逸生于地埂、沟渠、路旁,其枝叶无论干鲜各种家畜均喜食。

An annual herbaceous plant reproduced from seeds. It is usually cultivated in fields of over the northern China. The feeding value is high, the palatability is good, cattle and sheep and other animals will graze it.

六 荨麻科

9 大蝎子草

学名 *Girardinia palmata* (Forsk.) Gaud.
英名 Hemp-leaved Nettle

别名荨麻、麻叶荨麻。多年生草本。植株高50~120厘米,茎有棱,具螫毛。叶对生,具长柄,叶片掌状三深裂或三全裂,两面疏被柔毛,叶背还有螫毛。托叶离生,披针形或宽条形。花单性,雌雄同株或异株,同株者雄花序生于下方。花序长10~15厘米,多分枝。花被片4,雄蕊4。雌花被片4,深裂,花开后增大,柱头呈簇毛状。瘦果卵形,扁,灰褐色,光滑。在山地草原或

9-2 大蝎子草叶　　9-1 大蝎子草

戈壁边缘有时形成优势种分布区,早春返青较早,产草量高,对寒冷地区的放牧家畜具有重要的意义。骆驼贪食幼嫩青草过多会得"臌胀"致死,马、牛、羊在秋、冬两季最喜食。带籽喂鸡,能提高冬季产蛋率,喂猪增重快。

A perennial erect herb, reproduced by crown buds and by seeds. It is distributed in wet place and grasslands of dry or sand around Loess Plateau. The feeding value is moderate, and is very important forage for camels. Cattle and sheep like to graze it in early spring or after winter.

七 蓼科

10 旱型两栖蓼

学名 *Polygonum amphibium* L.
英名 Land Amphibious Knotweed

别名两栖蓼。多年生草本。根系发达,根节上可继续生根长芽。茎直立或斜升,基部分枝,被长硬毛。叶互生,具短柄。叶片披针形或宽披针形,先端急尖,基部近圆形,两面密生短硬毛,全缘;托叶鞘筒状。穗状花序顶生或腋生,花绿白色或淡红色。瘦果卵圆形,有钝棱,成熟时深褐色。主要着生在其他草较少的荒地、农田和果园。可以喂猪,经霜或干燥后家畜喜食。

A perennial herb, propagated mainly by rhizome and reproduced also by seeds. It occurs in farmlands and creek, with ability to grow in water, distributed in south and central Loess Plateau. It is harmful weed but a useful forage. Cattle and sheep like to graze after autumn frost. Pigs like it when boiled.

10-1 旱型两栖蓼　　10-2 旱型两栖蓼花序

11 东方蓼

11-2 东方蓼

11-1 东方蓼花序

学名 *Polygonum orientale* L.
英名 Prince's Feather

别名红蓼、水红蓼、大蓼、天蓼。一年生草本。茎直立,多分枝,密生长毛。叶互生,有长柄;叶片较大,卵形或宽卵形,先端渐尖,基部圆形或浅心形,全缘,两面疏生长毛;托叶鞘筒状或杯状。托叶下部膜质,褐色;上部草质,绿色。花穗长圆柱形,通常数个排列成圆锥状。花被5,深裂,淡红色。瘦果近圆形,扁平,两面微凹,先端有小柱状突起,黑褐色,有光泽。小麦、大豆、洋芋等作物中有见,幼苗或秋后家畜喜食,其他季节采食较少。

An annual herb reproduced from seeds. It is common existing in farmlands, watersides and wetlands as cultivated or volunteer plant, distributed throughout the country. The feeding value is moderate. Cattle and sheep like to graze when fresh.

12 酸模叶蓼

学名 *Polygonum lapathifolium* L.
英名 Dockleaved Knotweed

别名旱苗蓼、大马蓼。一年生草本。茎直立,有分枝,光滑无毛。叶互生,具柄。叶片披针形或宽披针形,叶面常有黑褐色新月或不规则斑块,无毛,全缘,边缘有粗硬毛;托叶鞘筒状,膜质。花序穗状,顶生或腋生,长圆柱状,通常数个排列成圆锥状。花被4,深裂,白色或淡红色。瘦果卵圆形,扁平,两面微凹,黑褐色。属北方常见草,多生于农田、果园和平坦低湿的地方。幼嫩枝叶可喂猪,其他家畜可食而不喜食,干燥后食性改善。

A annual herb, reproduced from seeds or by indefinite buds. It is commonly found in humid regions of mountain area, fieldsides and farmlands, distributed in south and central Loess Plateau. The feeding value is moderate. Cattle and sheep like to graze it when fresh or dry.

12 酸模叶蓼

13　绵毛酸模叶蓼

学名　*Polygonum lapathifoliun* L. var. *salicifolium* Sibth
英名　Hairy Garden Sorrel

别名白绒蓼。一年生草本。茎直立,有分枝。叶互生,具柄;叶片披针形或宽披针形,叶背密被白色绒毛层,叶面有或无黑褐色瓣块和毛;托叶鞘筒状,脉纹明显。圆锥花序由数个花穗组成;花淡红色或淡绿色。瘦果卵形,扁平,两面微凹。种子繁殖。鲜嫩多汁,作为青绿饲料羊、猪、鹅喜食,牛乐食,马不食。种子可作为多种畜禽和鸟类的精饲料。

13-1　绵毛酸模叶蓼　　　13-2　绵毛酸模叶蓼花序

An annual herbaceous plant reproduced from seeds. It is commonly found in roadsides, riverbeds and other moist lands. It is distributed throughout China. The feeding value is moderate with animals not grazing when it is fresh, but willing to graze it as dry.

14　萹　蓄

学名　*Polygonum aviculare* L.
英名　Common Knotgrass

别名鸟蓼、地蓼、扁竹、竹节草。一年生草本。茎自基部分枝,平卧或斜上,少有直立。叶互生,具短柄。叶片狭椭圆形或披针形,全缘,托叶鞘膜质。花1~5朵簇生于叶腋,全露或半露于托叶鞘之外。花被5,深裂,淡绿色,边缘白色或红色。瘦果卵状三棱形,深褐色上有不明显的小点,无光泽。多生于渠边、河边、路边或湿地。适口性好,牛、羊喜食。

An annual herb, reproduced from seeds. It occurs in roadsides, creek and farmlands, etc. It is distributed throughout the Loess Plateau except northern regions. It is a good forage, cattle, sheep and goats like it.

14-2　萹蓄花序　　　14-1　萹蓄

15　白花蓼

15-1　白花蓼

学名　*Polygonum coriarium* L.
英名　Whiteflower Knotweed

别名花蓼。多年生草本。株高1米，斜生或披散，茎稍有屈曲。叶片卵圆形或卵状披针形，先端渐尖，基部圆楔形，叶面绿色，叶背略带灰白呈浅绿色，叶脉清晰；叶鞘明显，暗紫褐色，被短柔毛。圆锥花序，多分枝，小花密布，花被白色。瘦果黄褐色。植株柔嫩多汁，营养价值比较高，青绿前期因有异味而家畜很少采食。8～9月结实后，特别是经秋霜以后，适口性极大改善，马、牛、羊乐食。

A perennial herb, spread by crown buds and by seeds. It occurs on steppes, meadow steppes and mountain forest sides, distributed in northwest China. Cattle and sheep like to graze after autumn frost.

15-2　白花蓼花序

16　西伯利亚蓼

16-2　西伯利亚蓼花序
16-1　西伯利亚蓼

学名　*Polygonum sibiricum* Laxm
英名　Siberian Knotweed

别名剪刀股。多年生草本。根茎和种子繁殖。通体无毛，根状茎细长。茎自基部分枝，斜上或近直立，高10～30厘米。叶互生，具短柄；叶片披针形、条形或矩圆形，近肉质，先端急尖，基部戟形或楔形，全缘，具腺点；托叶鞘筒状，膜质。圆锥花序顶生，苞漏斗状，内含5～6朵花，花被5，深裂，黄绿色。瘦果卵形，具三钝棱，黑色，有光泽。中等饲草，骆驼、山羊和绵羊乐食其嫩枝叶，牛、马不食，秋季经霜后适口性提高。

A perennial herbaceous plant propagated by rhizome and by seeds. It occurs in saline-alkali damp low lands, farmlands and wastelands. It is distributed in Gansu, shaanxi, Inner Mongolia and Sichuan etc. Livestock will graze after a frost in autumn.

17 尼泊尔蓼

学名　*Polygonum nepalense* Meisn
英名　Nepal Knotweed

别名野荞麦。一年生草本,种子繁殖。茎细弱,直立或匍匐,有分枝,高 30~50 厘米。叶互生,下部叶具长柄,常有狭翅,上部叶近无柄,抱茎;

17-1　尼伯尔蓼

叶片三角状卵形或卵形,先端渐尖,基部截形或圆形,有时呈耳垂形,全缘,叶背密生金黄色腺点;托叶鞘筒状,膜质。头状花序顶生或腋生;花被4,深裂,白色或淡红色。瘦果圆形至宽卵形,两面凸出,黑褐色,密生小点,无光泽。中等牧草,家畜季节性采食。

17-2　尼泊尔蓼花序

An annual herbaceous plant reproduced from seeds.It occurs in moist farmlands or water sides and is distributed in southeast Loess plateau as well as south of China.The feeding value is moderate and cattle and sheep will graze after an autumn frost.

18 珠芽蓼

学名　*Polygonum viviparum* L.
英名　Viviparous Bistort

多年生草本。须状茎肥厚,紫褐色;茎直立或斜升,不分枝,细弱,无毛,蔟生于根状茎上。基生叶有长柄;叶长圆形或披针形,革质,先端锐尖,基部圆形或楔形,边缘微向下卷;茎生叶有短柄或近无柄,披针形,较小;叶托鞘筒状,膜质。穗状花序顶生,圆柱形,中下部生珠芽;花淡红色,花被5,深裂。瘦果卵形,有三棱,深褐色,有光泽。是高山和亚高山草甸的主要植被之一,茎叶青鲜时山羊、绵羊乐食,马、牛可采食,骆驼不食;果实成熟后富含蛋白质,是家畜催肥抓膘的良好饲草。

A perennial herbaceous plant,spread by crown buds and by seeds.It is commonly found on mountain meadows in Tibet, Xinjiang,Qinghai,Gansu and Inner Mongolia.When it is fresh, sheep and goats will graze it.Horses,cattle and camel will graze a little,but it is a good forage for fattening in autumn.

18-1　珠芽蓼

18-2　珠芽蓼花序

19 柳叶刺蓼

19-1 柳叶刺蓼　　　　19-2 柳叶刺蓼花序

学名　*Polygonum bungeanum* Turcz
英名　Willowleaf Knotweed

别名本氏蓼。一年生草本。茎直立，多分枝，具倒生钩刺。叶互生，具短柄；叶片披针形或宽披针形，先端急尖，基部楔形，全缘，有睫毛；托叶鞘筒状，膜质，先端截形，有长睫毛。花序穗状，细长，花序轴密生腺毛；花排列稀疏；花被5深裂，白色或淡红色。瘦果近圆形，两面稍凸出，黑色，无光泽。种子繁殖。牛羊乐食。

An annual herbaceous plant reproduced from seeds.It occurs on wastelands,farmlands and grasslands.It is distributed in north and northeast China as well as other regions.The feeding value is moderate and most domestic animals will graze in spring and winter.

20 黏毛蓼

学名　*Polygonum viscosum* Buch
英名　Aromatic Knotweed

别名香蓼。一年生草本，种子繁殖。茎直立。上部多分枝，密生长毛；小枝具腺毛，微有黏性。叶互生，具柄；有狭翅；叶片披针形或宽披针形，全缘；两面疏生或密生糙伏毛；托叶鞘筒状，膜质，密生长毛，花序穗状。总花梗密生长毛或腺毛；花红色。瘦果卵状三棱形；黑褐色，有光泽。全国各地均有分布。中等饲草，青鲜时牛羊采食，蒸煮后可喂猪，秋后经霜家畜喜食。

An annual herbaceous plant reproduced from seeds.It occurs in water sides and wet lands of roadsides, distributed in Gansu,Shaanxi and northern China.Sheep,goats and camels will graze it after frost, but the pigs like to graze it after boiled.

20-2　黏毛蓼花序

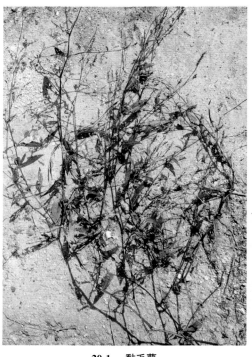

20-1　黏毛蓼

21 杠板归

学名　*Polygonum perfoliatum* L.
英名　Perfoliate Knotweed

别名犁头刺，蛇倒退。多年生蔓生草本。茎细长，有棱角和倒钩刺。叶互生，具长柄，密布倒钩刺；叶片三角形，叶面无毛，叶背沿叶脉疏生钩刺，托叶鞘叶状，近圆形，抱茎。短穗状花序，顶生或腋生；花被片5，深裂，白色或淡红色，结果时增大，肉质，成熟时变为深蓝色。瘦果球形，黑色，有光泽。生于沟壑底部较湿润的地方和林缘灌丛，家畜适口性差。

An perennial trailing herb, reproduced from seeds. It occurs in creek sides, wetlands and farmlands, grows among thickets in the valleys. It is distributed in Shaanxi and south, central China. The livestock animals like to graze young plant and dry leaves. It can be used for snakebites.

21-1　杠板归

21-2　杠板归花序

22 羊蹄

学名　*Rumex japonicus* Houtt.
英名　Japanese Dock

别名羊蹄叶，皱叶酸模。多年生草本。根系发达，主根粗壮。茎直立，通常不分枝。基生叶具长柄，叶片披针形或长圆状披针形，先端渐尖，基部楔形，边缘有波状皱折；茎生叶小而稀少，具短柄；托叶鞘筒状，膜质。圆锥状花序长而紧密，花簇轮生，繁盛，花两性；6花被片排成2轮，内花被片在果实增大时呈宽卵形，全缘或有齿，具网纹，有瘤状突起。瘦果椭圆形，褐色，有3棱。开春返青早，家畜喜食，夏秋时节适口性下降。

It is a perennial herb, reproducing from seeds and propagating by crown bud. It is commonly found in rather damp low field-sides, creek, and water sides. It is commonly distributed in Loess Plateau as well as many parts of Mongolia, Qinghai, Xinjiang, Fujian and Taiwan. The leaves can be fed for pigs, rabbits, cattle and sheep.

22-2　羊蹄　　　22-1　羊蹄花序

23 鲁梅克斯

学名 Rumex (R.Patientia × R.Tianschanicus Rumex K-1)

别名高秆菠菜、饲料酸模。多年生草本。根系粗壮发达，肉质，入土深。播种当年多为营养生长，第二年才开始生殖生长。茎直立，中下部具棱槽。基生叶卵状披针形，先端尖，基部圆，全缘，有长柄；茎生叶稀少，狭小，近无柄，托叶膜质。花序长而顶生，花两性，花量大，雌雄同株，玫瑰色。瘦果，锐三棱形，红褐色，有光泽。1997年引入甘肃省庆阳黄土高原试验站试种，喜水肥，能高产。适于青刈或青贮，不宜调制干草。青刈饲喂应在盛花期以前与其他牧草混合，单一饲喂适口性较差。

A perennial herb repro duced from seeds.It is suited to fertile and wet lands, and grows quickly in spring.The feeding value is moderate.It is palatable for cattle before flowering while sheep, goats and other animals graze limited.The green leaves can be fed for pigs.

23-2 鲁梅克斯

23-1 鲁梅克斯花序   23-3 鲁梅克斯根

八 藜科

24 灰绿碱蓬

学名 *Suaeda glauca* Bge
英名 Common seepweed

别名碱蓬。一年生草本，种子繁殖。茎直立，圆柱形，浅灰绿色，具纵条纹，上部分枝开展，可形成大丛。叶互生，肉质，无柄，条状丝形、半圆柱形或略扁平，先端钝或急尖，表面光滑或被粉粒，茎上部叶片渐短。花簇有短柄与叶的基部相连，单生或2~5朵排列成聚伞花序；花小，两性；球形；花被片5，果期增厚呈五角星状。胞果扁平，种子蜗牛状。多见于滨海地带或内陆河谷、荒漠草原的盐湿生境，属低等牧草，幼嫩时猪少食其叶，骆驼乐食，马、牛等大家畜一般不食。

An annual herbaceous plant reproduced from seeds.It is commonly found in saline-alkali farmlands, wastelands and canal banks.It is distributed throughout northeast and northwest China as well as Jiangsu and Zhejiang etc.The feeding value is low and camel will graze, horses and cattle do not graze.

24-2 灰绿碱蓬   24-1 灰绿碱蓬花序

25 绳虫实

学名 *Corispermum declinatum* Steph.exstev

英名 Chingan-mountain tickseed

一年生草本。茎多从基部分枝，直立或斜升。叶条形，先端渐尖，基部渐狭，有一脉。穗状花序顶生或侧生，上部稍紧密，下部疏松，苞片披针形至卵圆形，先端渐尖，具三脉，边缘膜质；花被片3；雄蕊5，长于花被片。胞果长圆状倒卵形或宽椭圆形，顶端圆形，基部心形，背面稍凸，腹面扁平，无毛，淡绿色，果翅明显。多生于撂荒地疏松的沙质土壤。适口性中等，青绿时骆驼采食，山羊和绵羊采食较少；干枯后骆驼喜食，马稍食，牛通常不食。

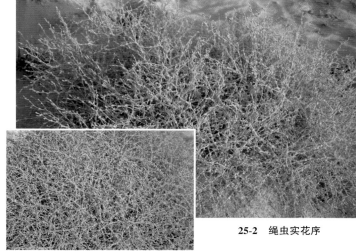

25-1 绳虫实

25-2 绳虫实花序

An annual herbaceous plant reproduced from seeds.It is commonly found in sandy lands and wastelands and is distributed in north and northwest China.The feeding value is moderate and camel will graze anytime, sheep and goats will graze when dry.

26 合头草

学名 *Sympegma regelii* Bunge

英名 Sympegma

小半灌木。茎直立，多分枝，老枝灰褐色，通常有条状裂纹；新生枝黑绿色。叶互生，肉质，圆柱形，黑绿色。花两性，常3~4朵聚集成顶生或腋生的小头状花序。花被片5，革质，果时变坚硬且自顶端横生翅。胞果扁圆形，果皮淡黄色。多生于石质荒漠，有时也出现在半荒漠地区，是骆驼主要的抓膘草之一，具有快速的催肥作用，羊在青鲜时稍食，干枯后喜食，马少食，牛不食。

26-1 合头草

26-2 合头草花序

A small shrub,commonly found on sandy loams and desert steppes.It is distributed in Xinjiang,Inner Mongolia and northwest Loess plateau.Camel will graze and it is an excellent forage for fatting.Cattle do not graze.

27　雾冰藜

27-1　雾冰藜

27-2　雾冰藜叶

学名　*Bassia dasyphylla*（Fisch. et Mey.）O. Kuntze.

英名　Fisch. et Mey

　　一年生草本，种子繁殖。茎直立，具条纹，多分枝；全株呈球形或卵状，密被水平伸展的白色长柔毛。叶互生，肉质，条状半圆柱形或圆柱形，先端钝圆，基部渐狭，无柄。花两性，每1~2朵集生于叶腋，仅一花发育；花被5，浅裂，圆壶形，具密毛；果期花被片背部生5个锥刺状附属物，呈五角星状。雄蕊5，伸出花被外。种子小，近圆形，光滑。散生于荒漠、半荒漠地区的固定或半固定沙丘、村落、畜圈附近，数量多，分布广。饲用价值不高，牛全年不食，马夏末秋初采食，山羊、绵羊秋季喜食，骆驼秋冬乐食。

　　An annual herbaceous plant reproduced from seeds. It occurs in desert steppes and sandy lands, and is found around sandy dunes, spread in north, northeast and northwest China. The feeding value is low and sheep and goats will graze in autumn, but cattle do not graze.

28　菊叶香藜

28-2　菊叶香藜花序　　　28-1　菊叶香藜

学名　*Chenopodium foetidum* Schrad

英名　Foetid goosefoot

　　一年生草本，种子繁殖。茎直立，分枝斜升，有条纹，通体疏生腺体或腺毛，有特殊气味。叶互生，具长柄；中、下部叶片长圆形，羽状浅裂或深裂，叶背疏生黄色粒状腺体和有节的短柔毛；上部的叶片渐小，浅裂或不裂。复二歧聚伞花序，多再次聚成塔形圆锥花序；花两性；花被片5，背部有刺突状的隆脊和腺点。种子扁球形，红褐色至黑色，有网纹。多生于农田和荒地，分布在"三北"地区和西南。中等饲用植物，适口性欠佳，秋冬季家畜采食较好，春夏季牛、羊乐食。

　　An annual herbaceous plant reproduced from seeds. It occurs in farmlands, wastelands, worn beaches and low hill slopes. It is distributed in north China and Loess plateau. The feeding value is moderate, horses, cattle and sheep will graze it all year around.

29　盐生草

学名　*Halogeton glomeratus*
英名　Beardless Halogeton

一年生草本。茎自基部分枝,近直立或斜升,枝条稠密,互生,表皮淡绿色,枝上无着生丝状毛。叶具易脱落的长刚毛,互生,肉质,圆柱形;叶腋生绵毛。花两性,通常2~3朵簇生于叶腋。5花被片在果期从背部生翅,翅半圆形,膜质,透明。胞果卵形,种子横生其中。遍布西北的干旱、盐碱、轻沙地区,耐寒、耐旱、耐盐碱,多生于山坡、沟谷、河滩及路渠两旁。骆驼喜食,幼嫩时其他放牧家畜也采食。

29-2　盐生草茎

29-1　盐生草

An annual herb,existing in sandy lands mountain areas river beaches,distributed in Xinjiang the western pan of Gansu Qinghai and Tibet.

30　白茎盐生草

学名　*Halogeton arachnoideus* Moq
英名　Halogeton

一年生草本。茎自基部分枝,枝条稠密,呈松散状斜升,互生,表皮白色,幼嫩时上面着生丝状毛(与盐生草的主要区别),后渐长渐落,变为光滑。叶互生,肉质,圆柱形,叶腋生绵毛。花两性,通常2~3朵簇生于叶腋。5花被片在果期从背部生翅,翅半圆形,膜质,透明。胞果卵形,种子横生其中。耐寒、耐旱、耐盐碱,喜生于沙地。骆驼喜食,幼嫩时其他放牧家畜也采食。

An annual herb.It commonly grows around sand-beaches and sandy lands in the desert.It is distributed over the Loess Plateau as well as many other provinces in northwest China,such as Xinjiang,Qinghai and Inner Mongolia.When it is fresh,camel will graze,but sheep and cattle graze limited.

30-1　白茎盐生草　　　　　　　　　　30-2　白茎盐生草茎

31 刺 藜

学名 *Chenopodium aristatum* L.
英名 Aristate Goosefoot

一年生草本,种子繁殖。茎直立,多分枝,高15~40厘米,有条纹,无毛或疏生柔毛。叶互生,叶柄不明显;叶片披针形或条形,全缘,两面均为绿色,秋季变为淡红色。复二歧聚伞花序,生于枝顶或叶腋,分枝末端有针刺状的不育枝。花两性;花被片5,绿色。胞果圆形,顶极压扁。种子圆形,边缘有棱,黑褐色,具光泽。生于田间、道旁、山坡及退化草地。属于中低等饲用植物,家畜采食较少。

An annual herbaceous plant reproduced from seeds.It is commonly found mainly in sandy farmlands and wastelands,and is distributed in north, northeast and northwest China as well as Shandong and Sichuan provinces.The feeding value is low and most animals graze a little.

31-1 刺藜

31-2 刺藜花序

32 尖头叶藜

学名 *Chenopodium acuminatum* Willd
英名 Acuminate Goosefoot

别名绿珠藜。一年生草本,种子繁殖。茎直立,多分枝,有条纹。叶互生,具短柄;叶片卵形或宽卵形,先端钝或急尖,具短尖头,全缘,叶面淡绿色,叶背有粉粒,灰白色。花序穗状或圆锥状,花序轴有毛;花两性;花被片5,果期背部增厚呈五角形状。胞果圆形,顶极压扁。种子扁圆形,黑褐色或黑色,有光泽。喜生于沙质农田、撂荒地、河滩或海滨,分布于我国"三北"地区、华东、及河南、广东诸省。饲用价值中等,猪喜食幼嫩部分,牛、羊在春季和秋后种子成熟时采食较好。

32-1 尖头叶藜花序

32-2 尖头叶藜叶

An annual herbaceous plant reproduced from seeds.It occurs in farmlands with sandy soil,river beaches and gullies.It is distributed in north,northwest and east China.The feeding value is moderate and livestock animals can consume a little in autumn.

33　灰绿藜

学名　*Chenopodium glaucum* L.
英名　Oakleaf Goosefoot

别名翻白藜,一年生草本。茎自基部分枝,枝条多平卧或上升,茎有绿色或紫红色条纹。叶互生,长圆状卵形至披针形,边缘有波状齿或近全缘。叶面深绿色,叶背灰白色,密生一层粉粒,叶片稍肉质加厚。花序穗状或复穗状,花两性或雌性,花被片3~4,仅基部合生。胞果扁圆形伸出花被外。种子扁圆形,红黑色或暗黑色。中上等饮用植物。

33-2　灰绿藜叶　　　　　　　　　　33-1　灰绿藜

An annual herb reproduced from seeds.It occurs in fields with moderate amounts of saline-alkali, roadsides and water sides.It is distributed in north, northwest and northeast China.The feeding value is high and livestock animals like to graze.

34　地　肤

学名　*Kochia scoparia*（*L*）Schrad
英名　Summer cypress

别名扫帚菜。一年生草本,种子繁殖。茎直立,高可达1米左右,多分枝,分枝斜生向上,呈扫帚状,具条纹,绿色或带淡红色。叶互生,披针形或条状披针形,扁平,先端渐尖,基部渐狭,无毛或稍生疏毛;叶面通常具3条纵脉,以中脉最明显。花两性或雌性,无柄,1~5朵花生于上部叶腋,于枝端排成穗状花序,疏密不等;花被片5,基部连合,黄绿色,卵形,内曲。胞果扁球形,包于花被内。种子横生,胚形。遍布全国各地,具有抗寒旱、耐风沙及适应盐碱地的特性。优良饮用植物,适口性好,草食家畜均喜食。

An annual herb reproduces from seed. It exists in farmlands, roadsides and wastelands, causes damages to melons and other dry-land crops and vegetables.It is distributed throughout the country. The feeding value is high and livestock aninnals like to graze.

34-1　地肤　　　　　　　　　　34-2　地肤叶

35 碱地肤

学名 *Kochia sieversiananus* (Pall.) C.A.Mey.
英名 Broomsedge

一年生草本。与地肤的主要区别在于：茎叶密被白绵毛，花1～2朵集生于叶腋的密毛中，多数花于枝上端排列成穗状花序。花被片5，果时花被片背部横生出5个圆形或椭圆形的短翅，翅具明显的脉纹，顶端边缘具钝圆齿。胞果扁球形，包于花被内。属于耐盐碱的旱生植物。幼嫩时人可食用，猪禽喜食。中等牧草，冬春季虽然有些叶子掉落，质地粗硬，但羊、驼依然嗜好。

An annual herbaceous plant reproduced from seeds. It occurs in roadsides and wastelands and is distributed throughout the country. The texture of the plant is soft, with a large amount of leaves. Various domestic animals like to graze it.

35-1 碱地肤　　　35-2 碱地肤花序

36 沙 蓬

36-1 沙蓬　　　36-2 沙蓬花序

学名 *Agriophyllum squarrosum* L.
英名 Agr iophyllum

别名沙米、登相子、蒺藜梗。一年生草本。幼嫩时通体密被分枝毛，后随生长逐渐脱落。茎直立，坚硬，多分枝。叶互生，无柄；披针形至条形，先端渐尖，有刺尖，基部渐狭，全缘，叶脉多条凸出。穗状花序，无总梗，通常1～3个着生于叶腋。苞片宽卵形，先端骤尖有短针刺，稍反折。花被片1～3，膜质。子房扁形，柱头2。胞果卵圆形，扁平，除基部外，周围略具翅，果喙深裂成2个条状小喙，其先端外侧各具一小刺。种子圆形，扁平。属耐旱植物，也是流沙上的先锋植物。在荒漠草原，骆驼终年喜食，是主要的催肥牧草。羊只采食嫩枝叶，马、牛不喜食。种子成熟度高，可作精饲料，是幼畜和怀孕母畜优良的补饲料。

An annual herb reproduced from seeds. It grows in sandy lands of field-sides and around sand dunes. It is distributed in north and northwest China. It is a pioneer plant of drift sand with high nutritive value. It becomes coarse and thorny after flowering. Camels like to graze in four seasons.

37　大苞滨藜

学名　*Atriplex centralasiatica* L.

一年生草本。株高20～50厘米，茎钝四棱形，具纵条纹，紫红色。由基部分枝，斜升，全株被白粉粒。叶互生，具短柄，叶片菱状卵形或卵状三角形，基部楔形或圆形，先端钝，边缘具不整齐的锯齿。小型叶片边缘波状齿不明显或近全缘，叶片两面密被白粉粒。花由2个合生苞片所包围，果时苞片膨大，呈倒卵形，具柄，中部两面凸。具多数棘状突起，顶端具齿，内含卵圆或近圆形胞果。种子扁球形，红褐或黄褐色。多生于荒漠草原或荒漠盐碱区。青绿时家畜不喜食，经霜后，牛、羊乐食。

37-2　大苞滨藜花序　　　37-1　大苞滨藜

An annual herb reproduced from seeds. It grows on alkaline or saline-alkali soils in hill slopes, valleys and sunny river banks. It is distributed in northwest China. The feeding value is moderate or low, cattle, sheep and camels like to graze after autumn frost.

38　中亚滨藜

学名　*Atriplex centralasiatica* Hjin.
英名　Central-Asia Saltbush

一年生草本。茎直立，多分枝，中部枝条发达。叶互生，三角状卵形或菱状卵形，边缘常有少数缺刻状锯齿，叶背密生粉粒。伞状花序遍生叶腋，花果簇生繁盛，枝条有负重感。花单性，雌雄同株。两雌苞片中部以下边缘合生，呈菱形或扇形，上部边缘有三角状齿。种子呈不规则的近椭圆形，土黄色。以盐碱地多见。家畜较喜食。

An annual herb reproduced from seeds. It is commonly found in saline-alkali farmlands, wastelands and river banks. It is distributed in north, northeast and northwest China as well as Tibetan region. When it is fresh, cattle, sheep, pigs and poultry like to graze.

38-1　中亚滨藜花序　　　38-2　中亚滨藜

39 鞑靼滨藜

学名 *Atriplex tatarica* L.
英名 Tartar Saltbush

一年生草本。茎由基部分枝，被白粉。单叶互生，具短柄，叶片稍呈宽卵形、三角形，基部广楔形，先端钝或急尖，边缘具不整齐的深波状钝齿，稀近全缘，表面绿色，背面灰白色。花单性雌雄同株，簇生于叶腋，成穗状，于茎上部构成穗状花序。雄花花被片5，雄蕊5，生于花托之上；雌花多数，聚成球状，具2苞，无花被，苞边缘合生直至顶部，卵圆形或近圆形，两面凸，膨大，成球状，果期木质化，包住果实，顶缘具牙齿，具多数棘状突起。中等牧草。青绿时适口性不高，秋霜后至渐干，牛、羊乐食。

An annual herb, it is growing on alkaline or saline-alkali soils in steppes, and distributed in north, northwest and northeast China. The feeding value is moderate and used by ruminants in autumn and winter and generally not in the spring and summer.

39-1 鞑靼滨藜　　39-2 鞑靼滨藜花序

40 盐爪爪

学名 *Kalidium foliatum*
英名 Kalidium

别名灰碱柴，半灌木。茎直立，斜升或平卧，多分枝，老枝灰褐色。叶互生，圆形，先端钝或稍尖，基部叶延伸半抱茎，肉质，灰绿色。穗状花序圆柱形或卵形，胞果圆形，红褐色，密被乳头状突起，种子和胞果同形。我国"三北"地区均有分布，生长在草原和荒漠的盐碱土壤上，散生或成群落状。冬春季节植株保存较好，对放牧具有重要意义。种子富含矿物质，与精饲料有同等的价值。

40-1 盐爪爪　　40-2 盐爪爪叶

A perennial salt-tolerant semi-shrub in sandy saline soils in desert region. It is distributed in northern Loess Plateau as well as northwest and northeast China. It is palatable for camels, but horses and sheep grazing is limited. In addition, the seeds can be edible, and can be used as animals feed in winter and spring.

41 刺沙蓬

学名 *Salsola ruthenica* Hjin.
英名 Russian thistle

别名刺蓬、苏联猪毛菜。一年生草本。茎直立或斜升,自基部分枝,坚硬,具白色或紫红色条纹。叶互生,条状圆柱形,肉质,先端有白色的硬刺尖。花1~2朵生于胞腋,通常在茎和枝的上端排列成为穗状花序。结果时由背侧中部横生5个翅,淡紫红色,苞果倒卵形。耐旱,喜疏松土壤,常常入侵农田。北方的多种放牧家畜采食,牛、马略差。

A strong xerophyte annual herb, it grows on migratory crescent dunes, on sandy hill lowlands over northern Loess Plateau. The seeds germinate quickly when it rains. It is a pioneer plant on drift sand-dunes with high nutritive value. Animals like to graze when fresh.

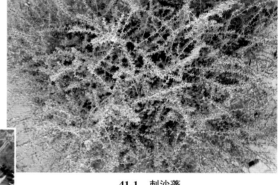

41-1 刺沙蓬

41-2 刺沙蓬花序

42 猪毛菜

学名 *Salsola collina* Pall.
英名 Common Russian-thistle

一年生草本。茎直立,高30~100厘米,基部多分枝,开展,有条纹。叶互生,条状圆柱形,肉质,具短糙硬毛,先端有小刺。花序穗状;苞片宽卵形,具硬刺,小苞片2,花被片5,披针形。胞果倒卵形。分布于东北、西北、四川、河南、江苏等地。

An annual herb, reproduces from seed. It exists in farmlands, roadsides and wastelands, damages wheat, cotton, peanuts, pulses and other dry-land crops. It is distributed in north and northeast China.

42-3 猪毛菜叶　　42-2 猪毛菜花序　　42-1 猪毛菜

43　兴安虫实

学名　*Corispermum chinganicum*
英名　Chingan-mountain tickseed

43-1　兴安虫实

43-2　兴安虫实花

别名绵蓬。一年生草本。茎直立，由基部分枝，下部分枝较长，上升。叶条形，先端渐尖，基部尖狭。穗状花序顶生或侧生，花序和叶片上部排列较紧密，下部疏松；苞片披针形至卵圆形，先端渐尖，边缘膜质。胞果长圆状倒卵形或宽椭圆形，顶端圆形，基部心形，背面稍凸，腹面扁平，无毛果翅。喜生于疏松土壤和沙壤，在干草原或撂荒地雨后能迅速生长，成为群落中的主要伴生种。通常用青干草或种子粉碎后补饲瘦弱病畜或幼畜。

An annual herb reproduced from seeds.It is commonly found in saline-alkali farmlands and moist saline-alkali soils, both as individual plants and in clump communities.This plant is widely scattered over the whole area of northern China.The feeding value is low with limited grazing vaule for sheep and goats while camels like the plant.

44　角果碱蓬

学名　*Suaeda cor niculata*（C.A.Mey.）Bunge
英名　Corniculate seepweed

一年生草本。无毛，茎平卧，外倾或直立，圆柱形，微弯曲，具细条棱，由基部分枝，枝斜升，稍弯曲，霜后呈紫红色，晚秋变黑。叶肉质，条形或半圆柱形，直或下部稍弯，先端微钝或锐尖，基部稍缢缩，无柄。团伞花序通常有花3~6，于分枝上排列成穗状花序；花两性，花被顶基略扁平，5深裂，裂片不等，先端钝，果时背面向外延伸增厚呈不等大的角状突起。种子横生或斜生，双凸镜形，黑色或黄褐色，表面具清晰的蜂窝状点纹。典型的盐生植物，分布于我国"三北"地区及西藏、内蒙古。低等牧草，青鲜时家畜不食，秋霜后植株变红，适口性大增。

44-1　角果碱蓬

44-2　角果碱蓬花序

An annual herb reproduced from seeds.It is commonly found in saline-alkali farmlands and moist saline-alkali soils.This plant is widely scattered over the whole area of northern China.The feeding value is low with limited grazing value for sheep and goats while camels like the plant.

45 小 藜

学名 *Chenopodium serotinum* L.
英名 Sall Goosefoot

一年生草本。茎直立,有分枝,具绿色条纹。叶互生,具柄。叶片长卵形或长圆形,边缘波状,具齿。中、下部叶片近基部有2个较大的裂片,叶两面疏生粉粒。花序穗状,腋生或顶生,花两性,花被片5,淡绿色。胞果包于花被内,果皮上有明显的蜂窝状网纹。种子双凸镜形,黑色,有光泽。多见于荒地、沟谷、河滩及农田,适口性好,家畜喜食。

An annual herb reproduced from seeds. It is commonly found in farmlands, river beaches, wastelands and humid river valleys, distributed throughout the country except Tibet. The feeding value is high. Pigs, sheep, rabbits and cattle like to graze it.

45-1 小藜

45-2 小藜花序

46 藜

学名 *Chenopodium album* L.
英名 Lambsquarters

别名灰菜、灰条、灰灰菜,幼嫩时人可实用。一年生草本。茎直立,高大,多分枝。下部分枝长,斜升或水平生长。叶互生,具长柄。基部叶片大,多呈菱形或三角状卵形,边缘有不整齐的浅裂或锯齿。茎上部叶片狭窄,全缘或具微齿,叶背面有粉粒。圆锥花序由多数花簇聚合而成,花两性。花被片5,胞果完全包被在花被内或顶端稍露。作为牧草,各种家畜喜食,是农区养猪的传统饲草,幼嫩时,反刍家畜采食量少。

An annual herb reproduced from seeds. It usually exists in wet and fertile lands, distributed throughout the China. It is forage with moderate feeding value. The seedling and seeds are quality feed for pigs and poultry.

46-1 藜   46-2 藜花序

47 杂配藜

学名 *Chenopodium hybridum* L.
英名 Mapleleaf goosefoot

47-1 杂配藜

47-2 杂配藜根系

一年生草本。茎直立，基部通常不分枝，无毛，有黄色或紫色条纹。叶互生，宽卵形或卵状三角形，先端急尖或渐尖，基部略呈心形、截形或近圆形，边缘裂深，脉纹清晰。圆锥花序顶生或腋生，花两性；花被片5，背面有纵向隆起。胞果双凸镜形，果皮膜质。种子黑褐色，无光泽，表面有明显的凹凸不平。幼嫩时家畜不喜食，干燥后可食。

An annual herb reproduced from seeds. It is commonly found in farmlands, roadsides, forest edges and canal sides, and distributed in north, northeast and northwest China as well as southern China such as Sichuan, Yunnan, etc. Cattle, sheep and goats graze in autumn and winter.

48 梭梭柴

学名 *Haloxylon ammodendron* (C. A. Mey.)
英名 Saxoul

多年生灌木或小半乔木。高度变化大。茎秆粗壮具节疣，扭曲，有纵向的条状凹陷，二年生枝条灰褐色。叶退化为鳞片状短三角形，基部宽，先端钝。花两性，单生于二年生枝条的叶腋；小苞片宽卵形，膜质，与花等长；花被5，矩圆形；果期自背部生翅状附属物，翅膜质半圆形，全缘，基部心形。种子黑色。生于半荒漠和荒漠地区，分布于甘肃、内蒙古西部，青海和新疆等地，是良好的饲用植物，骆驼最为喜食。

A perennial short semi-arborous, reproduces from seed. It exists in sandy saline and in the alluvial fans of mountains, dry river beds, lake basin lowlands and in gravel of the Gobi in Erdos.

48-2 梭梭柴   48-1 梭梭柴花序

苋 科

49 牛 膝

学名 *Achyranthes bidentata* Bl.
英名 Twotooth Achyranthes Root

多年生草本,种子和根芽繁殖。根细长,圆柱状,茎直立,四钝棱,高60~100厘米,节部膨大呈牛膝状,常带暗紫色。叶对生,具柄;叶片椭圆形至椭圆状披针形,先端尾尖,幼嫩时两面有柔毛。穗状花序顶生或腋生,花后总梗伸长,花向下折而贴近总花梗;苞片宽卵形,先端渐尖;小苞片2,刺状,基部有卵形膜质小裂片;花被片5,黄绿色。胞果长圆形,残存花柱长。中等牧草,放牧家畜喜食,尤其在秋季经霜以后,适口性改善。

A perennial herbaceous plant reproduced from seeds and propagated by crown buds. It occurs in farmlands, mountain slopes and wet lands of ditch sides and roadsides. It is distributed in Shanxi and the southern of Gansu as well as southern China.

49-1 牛膝

49-2 牛膝花序

50 籽粒苋

学名 *Amaranthus hypochondriacus* L.
英名 Prince's-feather

别名千穗谷。一年生草本,种子繁殖。茎直立,有钝棱,无毛或上部微有柔毛。单叶互生,倒卵形或卵状椭圆形,全缘或波状缘;叶柄绿色或略带淡紫色,无毛,无托叶。圆锥花序顶生,直立,圆柱形,中央穗序长度超过株高的1/3,由多数穗状花序组成,花簇在花序上排列紧密;苞片和小苞片卵状钻形,背部中脉隆起成长凸尖;花被片矩圆形,顶端尖,绿色或紫红色。开裂性周裂蒴果,种子近球形,白色、金黄色或黑色,其中黑色者有光泽。喜温植物,适宜半干旱半湿润地区。一种高产优质的饲草料作物,适口性好,马、牛、羊、猪、鸡、兔均喜食。

An annual herbaceous plant reproduced from seeds. It occurs in fairly moist and fertile farmlands, roadsides and home garden. It is distributed throughout the country. Palatability is good and nutritional value is high. Various domestic animals will graze it.

50-1 籽粒苋

50-2 籽粒苋花序

51 繁穗苋

学名 *Amaranthus paniculatus* L.
英名 Paniculate Amaranth

别名西黏谷。一年生草本，种子繁殖。茎直立，有沟棱，黄绿色或紫红色。叶互生，全缘，卵状椭圆形或卵状披针形，两面无毛；叶柄长，绿色，疏生柔毛。圆锥花序大，顶生或腋生，由多数穗状花序组成，直立，多分枝，顶生小花穗比侧生长；花单性，雌雄同株；苞片钻形，顶端成长芒；花被膜质。苞果卵形，盖裂，种子近球形，黄白色或棕褐色，有光泽。适应性强，耐酸、耐盐碱、耐瘠薄。畜禽的优质饲料，青饲、打浆、青贮、干草均可，各种家畜都喜食。

An annual herbaceous plant reproduced from seeds.It occurs in wastelands,home garden and orchards and is distributed over all parts of the country.It is a forage with high feeding value and has good palatability.Various domestic animals like to graze it.

51 繁穗苋

52 反枝苋

学名 *Amaranthus retroflexus* L.
英名 Shortbract Redroot Amaranth

别名苋菜、野苋菜。一年生草本。茎直立，分枝，密生短柔毛。叶互生有长柄；叶片卵形至椭圆状卵形，先端稍凸或略凹，有小芒尖，两面和边缘具柔毛。花序圆锥状，顶生或腋生，花簇刺毛多；花白色，5被片，具浅绿色中脉1条。胞果扁球形，包在花被中，开裂。种子圆形至倒卵形，表面黑色。喜生于水肥条件较好的农田、果园和荒地，主要分布在我国"三北"地区，其他地区也有分布。优良牧草，特别是幼嫩时期家畜喜食。籽粒也是家畜、家禽的精饲料。

An annual herb reproduced from seeds.It is commonly found in farmlands and wastelands, and distributed in north,northwest northeast China.It is an excellent forage and palatability is good.Cattle,sheep and goats like to graze its tender branches and seeds.

52 反枝苋

53　皱果苋

学名　*Amaranthus viridis* L.
英名　Wild Amaranth

别名绿苋、野苋。一年生草本，无毛。茎直立，有分枝，条纹明显。叶互生，叶片卵形至卵状长圆形，先端圆钝或微缺，有时具小芒尖，基部近截形，全缘，叶面常有"V"形白斑。圆锥花序顶生，是由花簇密集的细穗状花序组成的。苞片和小苞片干膜质。花被片3，长圆形或倒披针形，膜质。胞果扁球形，不开裂。种子圆形，略扁，黑褐色，有光泽。属优良牧草，家畜喜食。

An annual herb reproduced from seeds. It is commonly found nearby villages, wet-lands, road-sides and wastelands, and distributed in central and southern China. The feeding value is high and all animals like to graze it.

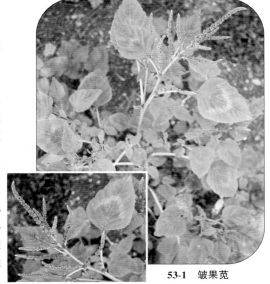

53-1　皱果苋

53-2　皱果苋花序

54　美国籽粒苋

学名　*Amaranthus hybridus* L.
英名　Hybrid Amaranth

别名绿穗苋、蛋白草、千穗谷等。一年生草本。为栽培历史悠久，形态多异，经济价值较高的饲料作物。其植株高大，产草量高，在庆阳黄土高原试验站栽培高度一般为1~1.5米，最高可达2米以上。生育期为110~140天，根系发达，根深50~80厘米。侧根多，根幅70~150厘米，茎粗3厘米。每667米^{2}鲜草产量高达6 000千克。老化前茎秆脆嫩，正常情况下茎叶生长繁茂，呈绿色或紫红色，营养物质含量高。无限穗状花序，花粉黄色，每叶基部都长有一侧枝，结一果穗。是牛、羊、猪、鸡、兔等畜禽的良好饲草饲料。

An annual cultivated herb reproduced from seeds. It is suited to fertile and wet lands. The yield is high and the feeding value is moderate with some domestic animals grazing before flowering. In addition, the seed is edible as quality feed.

54-1　美国籽粒苋　　54-2　美国籽粒苋花序

十　马齿苋科

55　马齿苋

学名　*Portulaca oleracea* L.
英名　Purslane

　　一年生草本。全株肉质、无毛。茎自基部分枝，平贴地面或仅先端斜上。叶互生或假对生，柄极短或近似于无柄。叶片倒卵形或楔状长圆形，全缘。花无梗，通常3～5朵簇生于枝顶。苞片4～5，膜质。花瓣5，黄色。蒴果圆锥形，盖裂。种子肾状卵形，稍扁，黑褐色，有小疣状突起。各种家畜均能采食，幼嫩时，人也可作为蔬菜食用。

　　An annual succulent herb reproduced from seeds. It is commonly found in wet lands and fertile lands. It is a terrible weed, especially in the lands of irrigation. The feeding value is high for pigs, sheep and other animals.

55-2　马齿苋

55-1　马齿苋花序

十一　石竹科

56　石　竹

学名　*Dianthus chinensis* L.
英名　China Pink

　　多年生草本。根粗壮，茎簇生，光滑无毛。叶对生，基部合生，抱茎。叶片条形或狭披针形，全缘。花单生、对生或呈聚伞花序，顶生于枝端。花萼合生，筒状。花瓣5，红色，紫色、粉红色或白色，极艳丽，边缘有不整齐的浅齿裂。蒴果长圆形；种子宽椭圆形至卵形，扁平，边缘有狭翅，灰黑色。主要分布在我国"三北"地区，生长于向阳山坡、路边、地边，适口性中等，春季幼嫩时家畜喜食。

　　A perennial herb, spread by seeds and by crown buds. It occurs in newly reclaimed farmlands, sunny hill slopes and field-sides, distributed in north, northeast and northwest China as well as around the Yangtze River valley. Sheep, cattle and goats like to graze the leaves.

56-2　石竹花序　　　　56-1　石竹

十一 石竹科

57 牛繁缕

学名 *Malachium aquaticum* L.
英名 Common Chickweed

一年生或二年生草本。茎自基部分枝，先端渐向上，茎贴地即生根，茎侧面有一行柔毛。叶对生，卵圆形，下部叶有柄，上部近无柄，先端尖锐，基部近心形，全缘。花单生于枝腋或顶端，萼片5，基部合生，花瓣5，白色，长于萼片，先端2深裂达基部。蒴果卵形或长圆形，5瓣裂。种子近圆形，略扁，深褐色，上面有不规则的小突起，种子和匍匐茎均可繁殖。茎叶柔嫩，适口性好，家畜喜食。

57-2 牛繁缕花序　　　　57-1 牛繁缕

An annual or biennial herb reproduced from seeds.It is commonly found in farmlands and orchards, and distributed throughout the country.When it is fresh, sheep and goats graze limited, but like to graze it after it is withered in autumn and winter.

58 瞿 麦

学名 *Dianthus superbus* L.
英名 Fringed Pink

茎圆柱形，上部有分枝，表面淡绿色或黄绿色，光滑无毛，节明显，略膨大，断面中空。叶对生，多皱缩，展平叶片呈条形至条状披针形。枝端具花及果实，花萼筒状；苞片4~6，宽卵形，长约为萼筒的1/4；花瓣棕紫色或棕黄色，卷曲，先端深裂成丝状。蒴果长筒形，与宿萼等长。种子细小，多数。生长于沟坡草丛及林缘，主要分布在黄土高原南部毗邻秦岭的地区。适口性中等，幼嫩或霜后家畜喜欢采食。

A perennial grass, it is commonly found on forest fringes, sparse woods, meadows and gullies.It is distributed over the entire country.The feeding value is low, and it can be used as a decorative plant.

58-1 瞿麦　　　　58-2 瞿麦花序

59 麦瓶草

学名　Silene conoidea L.
英名　Conical Catchfly

别名米瓦罐。越年生或一年生草本，种子繁殖。通体有腺毛，上部常分泌黏汁。茎直立，高25～60厘米，单生，叉状分枝，节部略膨大。叶对生，无柄，基部连合抱茎；基生叶匙形，茎生叶长圆形或披针形，全缘。聚伞花序顶生；花萼筒结果时呈圆锥状，有30条显著的脉棱；花瓣5，粉红色；花柱3。蒴果卵形，有光泽。种子肾形，扁，红褐色，有成行的瘤状突起。多生长在农田，分布于华北、西北至长江流域诸省。幼嫩时各种家畜都喜食，花后迅速粗老，适口性变差。

A biennial or annual herbaceous plant reproduced from seeds.It occurs in farmlands,damages winter wheat and other field crops.It is distributed in north and central China as well as Loess plateau.When it is fresh,cattle,sheep and goats will graze it.

59-1　麦瓶草

59-2　麦瓶草花序

60 女娄菜

学名　*Melandrium apricurn* Rohrb.
英名　Tatarinow Melandrium

越年生草本。主根圆锥形。茎直立或斜升，分枝稀少，多对生，密被腺毛。基生叶片大，莲座状，卵圆形或卵状长圆形；茎叶小，对生；叶缘具齿。花腋生，花萼具10条脉纹，花瓣成对，粉红色或粉白色，先端4裂。蒴果长卵圆形，种子少数，肾形，略扁。遍布"三北"地区，多生于海拔较高的山地、丘陵、林下，属于良好的饲用植物，通常用于放牧，各种家畜均采食，春季幼嫩时期和秋季经霜后适口性更好。

A biennial herb reproduced from seeds.It is commonly found in mountain areas or sandy-lands,distributed throughout the Loess Plateau.The feeding value is moderate.Horses,sheep and goats like to graze it in autumn.

60-1　女娄菜

60-2　女娄菜花序

61　蝇子草

学名　*Silene fortunei* Vis.
英名　Catchfly

多年生草本,种子繁殖。根圆柱形,粗而长。茎簇生,直立,有柔毛或近无毛,基部坚硬,节部膨大。叶对生,具短柄;基生叶匙状披针形,茎生叶条状披针形。聚伞花序顶生,总花梗上部有黏汁;花萼筒细管状,膜质,有10条脉纹;花瓣5,粉红色或白色,先端2深裂,裂片边缘不整齐;花柱3。蒴果长圆形,先端6齿裂。种子肾形,扁,有棒状突起,黄褐色。生于向阳山坡、地边、路旁或沟边草丛中,新垦地较多见。马、牛、羊均喜食其嫩枝叶。

A perennial herbaceous plant reproduced from seeds.It is commonly found in newly reclaimed farmlands, sunny hill slops and roadsides, and distributed around the Yangtze River valley and the southern part of the Loess plateau. Cattle, sheep and goats will graze its tender branches and leaves.

61-2　蝇子草花序　　**61-1　蝇子草**

十二　毛茛科

62　展枝唐松草

学名　*Thalictrum squarrosum* Steph.
英名　Nodding Meadowrue

别名歧序唐松草。多年生草本,根茎和种子均可繁殖。株高可达100厘米,通体无毛,叶集生于茎中下部,3~4回三出羽状复叶,小叶纸质,卵形或倒卵形,全缘或三浅裂,裂片钝或尖,有时具1~2锯齿,叶柄基部具短鞘;托叶膜质,撕裂状。圆锥花序二叉状分枝,开展。萼片椭圆形。瘦果,无柄,直或微弯,长圆状倒卵形。喜生于沙质撂荒地,干燥的砾质草地或林下,分布在黄河以北地区。春夏生长期枝叶较粗硬,含弱毒,秋霜后或调制成干草毒性就会消失,可与其他饲草配合饲喂家畜。

A perennial herbaceous plant reproduced from seeds and by crown buds.It is commonly found in mountain slopes and grasslands, distributed over the Loess plateau.It is a poisonous plant, but sheep and goats will graze after it is withered.

62　展枝唐松草

63　瓣蕊唐松草

学名　*Thalictrun potaloideum* L.
英名　Petalformed Meadowrve

多年生草本。无毛。分枝。3～4回三出复叶；小叶倒卵形、近圆形或菱形，3浅裂至3深裂，裂片卵形或倒卵形，全缘，脉平或微隆起；上面绿色，下面微带粉白色，有短柄。复单歧聚伞花序伞房状；萼片4，白色，卵形，早落，无花瓣；雄蕊多数，花丝倒披针形，比花药宽；心皮4～13，花柱短，柱头狭椭圆形。瘦果卵形，纵肋明显，宿存花柱直，长约1毫米。生于山坡灌丛和林缘草地中，分布西南、西北、华北和东北等地。干燥后家畜少食。

A perennial erect herb, usually found in Loess Plateau. It is a toxic plant for the animals. The livestock do not graze.

63-1　瓣蕊唐松草　　　　**63-2　瓣蕊唐松草花序**

64　东方铁线莲

学名　*Clematis orientalis* L.
英名　Oriental Clematis

半木质藤本，多年生。茎细长，茎、叶无毛。叶对生，羽状复叶，羽片有长柄。小叶灰绿色，狭卵形或披针形，全缘或具少量锯齿。单聚伞花序或圆锥状聚伞花序，花小，多少不一；萼片4，黄色或外带红紫色，内疏生柔毛；无花瓣；雄蕊多数，心皮多数，花丝有毛。瘦果椭圆形，有毛，上有宿存的花柱细长，羽毛状，弯曲。零星或成丛生长在干旱或较干旱山坡、丘陵、沙地，广泛分布于西北诸省。开花结籽以前幼嫩茎叶是家畜的好饲草，山羊尤为喜食。

A semi-ligniferous liana, it is commonly found in orchards and among thickets in mountain valleys, and distributed in Loess Plateau and Xinjiang. The feeding value is high and livestock animals will graze it especially goats when it is green.

64-1　东方铁线莲

64-2　东方铁线莲花序

65　甘青铁线莲

学名　*Clematis tangutica* (Maxim) Korsh.
英名　all-grass of Longsepal Clematis

落叶藤本。主根粗壮,木质。茎具棱,幼时被长柔毛,后脱落。叶对生,一回羽状复叶;叶柄长,叶片浅裂、深裂或全裂,狭长圆形或披针形,先端钝,有短尖头,基部楔形,边缘有不整齐缺刻状锯齿。花单生,有时为单聚伞花序,有3朵花,腋生;花序梗粗壮,有毛;花两性;萼片4,狭卵形、椭圆状长圆形,黄色外面带紫色,斜上展;花瓣无;雄蕊多数,花丝下面稍扁平,被开展的柔毛;心皮多数,密生柔毛。瘦果倒卵形,有长柔毛。主要分布在甘肃、青海、陕西、新疆和西藏的天然草原。

65-1　甘青铁线莲

A semi-ligniferous liana, commonly found in newly reclaimed farmlands, forest edges and among thickets in mountain valleys, distributed in Gansu, Qinghai, Xinjiang and Shaanxi provinces, etc. The livestock do not graze.

65-2　甘青铁线莲花序

66　大火草

学名　*Anemone tomentosa* (Maxim) Pei.
英名　Tomentose Windflower

别名野棉花。多年生草本。根粗大,入土深。基生叶3~4片,三出复叶,小叶卵形,三裂,边缘有粗锯齿。叶面绿色,被短伏毛,背面灰色,密被白色绵毛。聚伞花序,花无瓣,5萼片,粉红色,倒卵形,雄蕊多数,花丝线状,子房被柔毛。根有毒。茎叶经霜后,家畜采食,制成干草或草粉饲喂效果更好。

A perennial herb that grows in hill slopes, and distributed in Shanxi, Gansu, Sichuan, Henan and Hebei provinces, etc. The feeding value is low and most animals do not like it. The pigs can eat its leaves as powder.

66-1　大火草

66-2　大火草花序

67 升 麻

学名 *Cimicifuga foetida* L.
英名 Shunk Bugbane Rhizome

别名绿升麻、西升麻、川升麻。多年生草本。茎直立,高1~2米,上部分枝。下部茎生叶具有长柄,叶片三角形或菱形,2~3回三出羽状全裂,叶下表面沿脉疏被白茸毛或全部被白色绵毛;茎上部的叶较小,具短柄或近无柄,常1~2回3出或羽状全裂。圆锥花序,花序轴和花梗密被灰色或锈色的腺毛或短毛;花两性;萼片5,花瓣状;雄花多数。骨突果长圆形,被贴伏的柔毛,顶端有短喙。种子具膜质鳞翅。生于水热条件较好的山地林缘、林中或路旁草丛中。

67-1 升麻花序　　　67-2 升麻

A perennial herbaceous plant reproduced from seeds. It occurs in farmlands of mountain areas, hill slops, forest edges and among shrubs. It is distributed in Qinghai, Sichuan, Shaanxi and the southern part of lower and middle region of Gansu provinces.

68 绵团铁线莲

学名 *Clematis hexapetala* Pall.
英名 Sixpetal Clematis Root

多年生直立草本。高30～100厘米,根呈圆柱形,表面棕褐色,质坚脆,易折断,根茎呈不规则圆柱形。茎有纵棱,疏生短毛。羽状复叶对生,中部及下部的小叶常羽状全裂,裂片革质,通常条形。聚伞花序腋生或顶生,具3朵花,苞片条状披针形,花梗有伸展的柔毛;萼片6,白色,具白色柔毛,无花瓣。瘦果倒卵形、扁。生于山地、林边或草坡上,主要分布在东北和山东。家畜可采食嫩枝和叶片。

Perennial herbaceous plant, commonly found in among hassocks along hill-slopes and forest edges. It is distributed in Shandong and northeast China as well as southeast of the Loess plateau. Livestock animals will graze its fresh leaves.

68-2 绵团铁线莲花序　　　68-1 绵团铁线莲

十三 罂粟科

69 角茴香

学名 *Hypecoum erectum* L.
英名 Herb of Thinfruit Hypecoum

一年生或越年生草本，种子繁殖。通体无毛有白粉。1～10条茎抽生于叶丛中，细弱，有分枝，高10～30厘米。基生叶多数，2～3回羽状全裂，小裂片纤细，条形；茎生叶细小或无。聚伞花序具少数或多数分枝；萼片2，早落；花瓣4，黄色，外面2瓣较大，扇状倒卵形，里面2瓣较小，楔形；雄蕊4，蒴果条形。种子深褐色至黑色，有棱状突起。生于干燥的山坡草地、沙荒地，分布于西北、华北。家畜多不采食或经霜半枯后少量采食。

69-1 角茴香

69-2 角茴香花序

A biennial or biennial herbaceous plant reproduced from seeds. It occurs in grasslands, farmlands and sandy wastelands. It is distributed in north China as well as Gansu, Shaanxi, Henan and Xinjiang, etc. Usually, livestock do not graze it.

70 野罂粟

70 野罂粟

学名 *Papaver nudicaule* L.
英名 Nakestem Poppy

别名野大烟、山米壳。多年生草本，种子繁殖。具白色乳汁，全株有硬伏毛。叶基生，有长柄，两侧具狭翅，被刚毛。叶片卵形或窄卵形，1回深裂，最终小裂片狭矩圆形、披针形或狭三角形，两面被刚毛；花葶一至多条，圆柱形。单花顶生，稍下垂。萼片2，早落；花瓣4片，外2片较大，内2片较小，倒卵形，花色多样，以淡红、粉红、橘黄色为多见。蒴果卵圆形，顶孔裂。种子多数，细小、黑色。多生长在海拔高、光照好、土壤潮湿的荒草地，经霜或干燥后有的家畜少量采食。

A perennial herbaceous plant reproduced from seeds. It occurs on alpine grasslands and alpine meadow, grassy hill-slopes or among shrubs of forest edges. It is distributed between the Loess plateau and Qinghai-Xizang plateau.

71 多刺绿绒蒿

学名 *Meconopsis horridula*
英名 Spiny Meconopsis

一年生草本。全体被黄褐色或淡黄色坚硬而平展的刺。主根肥而长。茎近无或极短。叶均基生，披针形，两面被上述的刺；花葶坚挺，常5~12条，密被黄褐色平展的刺。花单生于花葶顶，半下垂，花瓣5~8，有时为4，蓝色或蓝紫色，宽倒卵形。蒴果倒卵形或椭圆状长圆形。生于高海拔地区的山坡、石缝之中，甘肃、青海辖区的青藏高原较为多见。

An annual herbaceous plant reproduced from seeds.It is commonly found on alpine grasslands and mea- dow steppes of rocky remain.It is distributed between the Loess plateau and Qinghai-Xizang plateau.Sheep will graze it in autumn.

71 多刺绿绒蒿

十四 十字花科

72 荠 菜

学名 *Capsella bursa-pastoris* L.
英名 Shepherdspurse

别名荠荠菜。一年生或越年生草本。茎直立，有分枝，全株被分枝毛或单毛。基部叶丛生，大头羽状分裂，边缘有不规则的分裂，叶具柄。茎生叶披针形，基部抱茎，边缘有锯齿或缺刻。总状花序顶生或腋生，花瓣4，白色。短角果倒三角形或倒心形，扁平，先端微凹。种子椭圆形，黄褐色。生于农田、果园、村庄、道路以及休闲地、免耕地、撂荒地。幼嫩时人可作为野菜食用，马、牛、羊、猪、兔等均喜食，鲜草有辛辣味，有些家畜不喜欢连续长时间采食，干燥后不失为牛、羊优质饲草。

72-1 荠菜

A biennial or annual herb reproduced from seeds.It is commonly found in farmlands and roadsides, distributed throughout the country.When it is fresh, the leaves are edible, and most animals like to graze this plant.

72-2 荠菜花序

73 离子芥

学名　*Chorispora tenella* (Pall.) DC.
英名　Tender Chorisporab

一年生草本。疏生短腺毛。茎斜上或铺散，有分枝。基生叶和下部叶羽状浅裂，长椭圆形或矩圆形，中部叶有短柄，上部叶无柄，披针形，具疏齿或全缘。总状花序顶生，花紫色，十字形；长角果细圆柱形，长3～5厘米，弯曲，具横节，不裂，先端有长喙。种子椭圆形，淡褐色。中旱生植物，能充分利用早春融化的雪水或降雨迅速生长，开花前草质柔嫩，适口性尚好，家畜采食。花后易粗老，除羊少食外，马、牛等大家畜通常不食，晒干后家畜在冬春季喜食。

73-2　离子芥花序　　　73-1　离子芥

A annual herb reproduced from seeds.It is mainly distributed around Loess plateau, major weed in small grain crops and winter wheat.The feeding value is low, and livestock like to graze this plant as hay.

74 弯曲碎米荠

学名　*Cardamine flexuosa* With
英名　Flexuose Bittercress

一年生或越年生草本。茎自基部分枝多，枝条斜升或呈铺散状，通体被疏柔毛。羽状复叶不整齐，小叶多形，有卵形、倒卵形、条形或长圆形等，先端稍裂或全缘，有小叶柄或无。总状花序多数，顶生。花瓣4，白色。细条形长角果，扁平，果穗轴有小弯曲。种子长圆形，黄绿色，先端有极狭的翅。喜生于潮湿的土壤，家畜喜食。

74-1　弯曲碎米荠　　　74-2　弯曲碎米荠花序

A biennial or annual herb reproduced from seeds.It is commonly found in wastelands, farmlands and grasslands, distributed in south to the Yangtze River as well as northern China, such as Shaanxi and Gansu.Cattle, sheep, goats and pigs like to graze in spring and autumn.

75 播娘蒿

学名 *Descurainia Sophia* L.
英名 Flix Weed Tansymustard

越年生或一年生草本。通体有分叉毛；茎直立，高大，上部多分枝。叶互生，上部叶无柄，下部叶有柄；叶片2~3回羽状深裂，小裂片窄条形或条状长圆形。总状花序顶生，有花多数。萼片4，直立。花瓣4，黄色。长角果细条形，有较长的柄，斜上伸展，成熟时容易开裂。种子细小，卵形或长圆形，黄褐色。小麦、油菜、果园地、路边、渠边及山坡草地最为多见。早春幼苗时家畜少量采食，一旦粗老后适口性迅速下降，基本不食。

A biennial or annual herb reproduced from seeds. It is commonly found in farmlands, wastelands and other dry-land crops, and distributed over Loess Plateau and eastern China as well as southern China such as Sichuan province. The feeding value is low. Animals like to graze the young plant as hay in winter.

75-1 播娘蒿

75-2 播娘蒿苗

75-3 播娘蒿花序

76 小果亚麻芥

学名 *Camelina microcarpa* L.
英名 Camelina

一年生草本。茎直立，不分枝或少分枝，被硬毛和分叉毛，上部近无毛。上部叶片披针形，先端锐尖，基部箭形，全缘，无柄。茎下部叶片大，矩圆状倒卵形。总状花序顶生，长。萼片4，矩圆状披针形。花瓣4，黄色，条形。短角果多数，有短柄，扁球形或梨形，光亮，先端具宿存的花柱。种子椭圆形，红褐色。现蕾前家畜采食，粗老后采食量减少，属于低等牧草。晒干可作冬季补饲牧草，家畜喜食。

A annual herb, reproduced from seeds. It grows on hill slopes and farmland and clay soils of desert regions. It is important forage in early spring, livestock animals like to graze before flowering.

76-1 小果亚麻芥   76-2 小果亚麻芥花序

77 遏蓝菜

学名 *Thlaspi arvense* L.
英名 Bastard Cress Beesomweed Pennycress

越年生或一年生草本,种子繁殖。通体无毛。茎直立,高15~40厘米,有棱,不分枝或少分枝。基生叶有柄,倒卵状矩圆形,全缘,枯萎早;茎生叶无柄,长圆状披针形,基部两侧箭形抱茎,边缘具疏齿。总状花序顶生;花瓣4,白色。短角果倒卵形或近圆形,扁平,边缘有翅。种子倒卵形,黄褐色,粗糙。多生于沟底、路边、农田及潮湿的地方,生活力极强,分布较广,我国"三北"地区、西南、江苏均有分布。青绿时牛、羊稍食,其他家畜不采食。

An annual or biennial herbaceous plant reproduced from seeds.It occurs in farmlands, wastelands, kill slopes and around gullies, and is distributed in Loess plateau, northeast and southwest China.Cattle sheep and pigs will graze it before flowering in spring or summer.

77-1 遏蓝菜　　77-2 遏蓝菜花序

78 独行菜

学名 *Lepidium apetalum* Willd
英名 Peppergrass, Pepperweed

越年生或一年生草本。茎直立或斜升、披散,多分枝。基生叶丛生,具柄,羽状浅裂或深裂;茎生叶互生,无柄;叶片条形,全缘或有疏齿。总状花序顶生,花小;萼片早落;花瓣退化为丝状。角果宽楔形,有柄,顶端中有凹缺,上部有极狭翅。种子倒卵状椭圆形,棕红色。遍布"三北"地区,多生于路旁及土壤紧实的地方。家畜较喜食,只是鲜草具辛辣味,家畜不喜欢连续采食。

A biennial or annual herb reproduced from seeds.It occurs in roadsides, farmlands and orchards.It is distributed in north, northeast, northwest and southwest China.The feeding value is low and most animals are readily to graze this plant when it is dry.

78-1 独行菜　　78-2 独行菜花序

79 宽叶独行菜

学名 *Lepidium latifolium* L.var.
英名 Grande Passerage

别名大辣辣。多年生草本。主根粗壮发达,基部分枝,茎直立,茎上部分枝短圆状披针形或卵形,先端急尖,基部楔形,具齿或全缘;上部叶卵状披针形,无柄。总状花序顶生,呈圆锥状。花瓣白色,稍向后卷。短角果宽卵形或近圆形,无毛,种子椭圆形或长圆形。多生于路旁、田埂、沙滩,有时也入侵农田和果园。家畜喜食幼苗和嫩枝叶。

79-1 宽叶独行菜　　79-2 宽叶独行菜花序

A perennial herb that spreads by seeds and by crown buds. It is commonly found in farmlands, roadsides and sand beaches. It is distributed in northwest China. The feeding value is low. Livestock graze a little when it is tender, but like to graze after a frost in autumn.

80 风花菜

学名 *Rorippa palustris* (Leyss.) Bess.
英名 Bog Marshcress

越年生草本。茎直立,有分枝。基生叶和茎下部的叶羽状分裂,顶叶较大,卵形,侧生裂叶片较小且边缘有钝齿,茎生叶愈向上愈小,分裂或不分裂。总状花序顶生或腋生,花瓣黄色。长角果圆柱状长椭圆形,稍弯曲,先端急尖如喙。种子卵形,稍扁平,淡黄褐色。鲜草家畜可采食,调制成干草采食率更高,马、牛、羊冬季均喜食。

A biennial herb reprod uced from seeds. It is commonly found in damp farmlands, roadsides and creek. It is distributed in provinces of north, northeast and northwest China as well as Jiangsu province. The palatability of dry plant is better than fresh. Cattle, sheep and goats like to graze it.

80-2 风花菜花序　　80-1 风花菜

81 串珠芥

学名　*Torularia humilis*
英名　Glabrous Low Torularia

别名蚓果芥、念珠。一年生或二年生草本。基部多分枝，斜升或铺散，具毛或分叉毛。叶片矩圆形或倒卵形，全缘或具齿。基生叶较大，莲座状，下部叶柄长。总状花序密集，果期伸长。萼片矩圆形，背面具单毛或分叉毛。花冠十字形，花瓣4，初开时白色，后逐渐变为粉红色，先端钝圆，基部渐狭成爪。长角果条形，直或弯曲，先端具短喙，有毛。多生于海拔1 000～2 000米的森林草原和典型草原，农区休闲地、果园亦有。青鲜时家畜采食，结实后蛋白质含量明显降低，粗纤维素增加，适口性随之下降。

81-1　串珠芥

81-2　串珠芥花序

A biennial or annual herb reproduced from seeds.It grows on shady side of the hills, grasslands and under forest and is distributed throughout the Loess Plateau.The feeding value is low.Goats and sheep like to graze, and the horses, cattle and rabbits graze limited.

82 垂果蒜芥

学名　*Sisymbrium heteromallum* C. A. Mey.
英名　Droopingfruit Garliccress

越年生或一年生草本。茎直立，上部有分枝，具粗硬毛。叶互生，中、下部叶有柄，被毛，叶片长圆形或长圆状披针形，大头羽状分裂，裂片2～4对，顶生裂片大，长圆状卵形，有不等的微齿，侧生裂片长圆形，有疏齿或近全缘；上部叶无柄，羽状浅裂，裂片披针形或宽条形。总状花序顶生；花小、淡黄色。长角果条形，无毛；果梗弯曲或开展。种子宽椭圆形或近方形，黄褐色。种子繁殖幼嫩时牛、羊喜食，开花后迅速粗老，适口性急剧下降。

82-2　垂果蒜芥叶

A biennial or annual herb, reproduces from seed.It exists in damp farmlands, hill slopes and plain lands, damages wheat and other dry-land crops.It is distributed in north and northeast China as well as Gansu and Shaanxi.The feeding value is moderate.

82-1　垂果蒜芥

83　桂竹糖芥

学名　*Erysimum cheiranthoides* L.
英名　Wormseed Sugarmustard

83-2　桂竹糖芥花序

别名小花糖芥。越年生或一年生草本。全体有伏生的2~4叉状毛。茎直立，有条棱。上部有分枝。叶互生，基生叶与下部叶有柄，叶片披针形、狭椭圆形或条形，边缘疏生深波状齿或羽状浅裂；上部叶片小，具波状齿或近全缘，无柄。总状花序顶生；花瓣4，淡黄色，长角果条状四棱形，略弯曲。种子椭圆状卵形至长卵形或近长圆形，黄色或黄褐色。种子繁殖。春季幼嫩时家畜少量采食，调制成青干草在枯草季饲喂牛、羊等，适口性好，采食率高。

83-1　桂竹糖芥

A biennial or annual herb, reprodces from seed. It occurs in dry hill slopes, commonly found in farmlands and roadsides. It is mainly distributed all over the country except south China. Sheep and cattle like to graze fresh branches and leaves.

84　垂果南芥

学名　*Arabis pendula* L.
英名　Pendentfruit Rockcress

越年生或多年生草本，种子繁殖。茎叶疏生硬毛和星状毛。茎直立，高20~80厘米，基部木质，不分枝或少分枝。中下部叶有柄，叶片长圆形或长圆状卵形，基部窄耳状，稍抱茎，边缘具疏齿；上部叶狭椭圆形或披针形，无柄。总状花序顶生，花白色，条形。长角果，扁平，下垂，具1脉。种子卵形，淡褐色，具狭边。常生于林缘灌丛、沙质草地，分布在"三北"及西南各地。富含粗蛋白质和灰分，适口性较好，家畜较喜食。

84-2　垂果南芥果穗

A biennial or perennial herbaceous plant reproduced from seeds. It occurs in farmlands, wastelands and grows among shrubs along forest edges. It is distributed in north, northwest, northeast and southwest China. The feeding value is moderate and animals will graze it.

84-1　垂果南芥

十五 景天科

85 费 菜

学名 *Sedum aizoon* L.
英名 Aizoon Stonecrop

别名景天三七。多年生草本。旱中生植物，主要依靠种子繁殖。茎直立，通体无毛，多丛生。叶互生，近无柄；叶片长披针形至倒披针形，边缘中、上部有锯齿，下部全缘。聚伞花序顶生；花瓣5，黄色；雄蕊10，短于花瓣；心皮5，基部合生。蓇葖果星芒状排列，叉开呈水平状。

A perennial herbaceous plant.It is commonly found in fieldsides and grasslands, grows among hassocks along hill slopes, or in the cracks between rocks around gullies.It is distributed in north China.and around the Yangtze River valley.

85-2 费菜花序

85-1 费菜

十六 鼠李科

86 酸 枣

学名 *Ziziphus jujuba* var. *spinosa*（Bumge）Hu.
英名 Spine Date

落叶灌木，稀乔木。皮褐色或灰褐色，枝条有长枝、短枝和脱落枝之分。长枝舒展，呈"之"字形折曲，红褐色，光滑，有托叶刺；短枝通常称枣股，在二年生以上的长枝上互生，似长乳头状；脱落性小枝亦称枣吊，为纤细下垂的无芽枝，似羽叶的总柄，常3~7簇生在短枝节上。叶互生，椭圆状卵形、卵形或椭圆状披针形，先端钝尖，基部稍偏斜，基生3主脉，具钝锯齿，两面光滑，有柄。花黄绿色，两性，单生或2~8朵密集成腋生的聚伞花序。花瓣倒宽卵形，基部有爪。核果近球形，成熟时红色，果肉酸味，核两端钝。果实和幼嫩枝叶营养丰富，多种草食家畜，特别是山羊最喜食，干燥后脱落的叶子也是放牧家畜冬春的优质饲料。

An erect shrub, it grows on slopes, field-sides and precipice edges. It is distributed throughout the Loess Plateau.Sheep and goats will graze its leaves.The nutritive value is high and palatability is very good in winter.

86 酸枣

十七 蔷薇科

87 蛇莓

学名 *Duchesnea indica* Focke
英名 Indian Mockstrawberry

多年生草本。具细长匍匐茎,有柔毛。三出复叶,小叶菱状卵形或卵形,边缘有钝锯齿,两面散生柔毛。叶柄长,托叶卵状披针形。花单生于叶腋,花梗长,有柔毛。花托扁平,果期膨大成半球形,红色,海绵质。花瓣5,黄色。花萼裂片卵状披针形,副萼片略大于萼片,5枚,先端3~5裂,均有柔毛。瘦果长圆状卵形,暗红色。生于较潮湿的红土沟壑下部草丛。家畜采食而不喜食。

87-2 蛇莓果　　　　**87-1 蛇莓**

A perennial herb, it is spread by seed and by stolons.It is commonly found in wastelands, creek sides, commonly found in hill-slopes among hassocks.It is distributed all over the country except northern part of the Loess Plateau.Sheep and goats will graze the leaves in winter.

88 委陵菜

学名 *Potentilla chinensis* L.
英名 Chinese Cinquefoil

别名翻白菜、老鸦翎、白头翁。多年生草本。株高30~60厘米,根圆柱状,木质化,黑褐色。茎直立或斜升,有毛。奇数羽状复叶,基生叶丛生,具10~25片小叶,小叶椭圆形或长椭圆形,羽状中裂或深裂,裂片三角状披针形,下面密生白色的柔毛和毡毛。茎生叶与基生叶相似。伞房状聚伞花序顶生,花多数,黄色。瘦果肾状卵形,略有皱纹。各种家畜均喜食,以羊为最。

88-2 委陵菜花序　　　　**88-1 委陵菜**

A perennial herb, spread by seeds and by crown buds.It occurs in farmlands of mountain areas, hill slopes, creek sides and nursery gardens, is distributed in south and central Loess Plateau.The feeding value is high.Horses, cattle sheep and goats like to graze it especially in autumn.

十七 蔷薇科

89　匍枝委陵菜

学名　*Potentilla flagellaris* Willd.
英名　Runnery Cinquefoil

多年生草本。匍匐茎,细长,幼嫩时被长绒毛,以后渐长渐脱落。基生叶为掌状复叶,小叶5,长圆状披针形,边缘有不规则的浅裂,叶柄长,叶背脉和叶柄均有柔毛,茎生叶稀而小。花单生于叶腋,花梗细长,有柔毛。花瓣5,黄色。瘦果长圆形,微皱,疏生柔毛。遍布黄土高原,多生于河边、路旁、草甸,优良牧草,家畜喜食。

89-2　匍枝委陵菜不定根

89-1　葡枝委陵菜

89-3　葡枝委陵菜花序

A perennial herb, spread by stolon and by seed. It is commonly found in meadows, river banks and nursery gardens, distributed throughout the Loess Plateau. The feeding value is high. Cattle, sheep and goats like to graze in spring and autumn.

90　朝天委陵菜

学名　*Potentilla supina* Linn.
英名　Carpet Cinquefoil

一年生或越年生草本。茎自基部分枝,茎较其他委陵菜粗壮,直立或斜升,疏生柔毛。羽状复叶,小叶长圆形或倒卵形,边缘有缺刻状锯齿,叶面无毛,叶背少毛。茎生叶有时为三出复叶,托叶宽卵形,三浅裂。花单生于叶腋,花梗细长,有柔毛。花瓣5,黄色。瘦果卵形,黄褐色,有皱纹。西北有零星分布,多生于潮湿的农田、果园,家畜喜食。

A biennial or annual herb reproduced from seeds. It is commonly found in roadsides and water sides, distributed in north, northeast and northwest China as well as southern China such as Hunan, Hubei and Anhui. The feeding value is moderate. Cattle and horses will graze when fresh and sheep like to graze it in autumn.

90-1　朝天委陵菜

90-2　朝天委陵菜花序

91　二裂委陵菜

学名　*Potentilla bifurca* L.
英名　Bifurcate cinquefoil

多年生草本。单数羽状复叶，有小叶5～13片，椭圆形或倒卵状椭圆形，侧生小叶顶端常二裂，叶柄有毛。伞房状聚伞花序生于枝顶，花黄色，瘦果稀少。常生于干旱山坡、低埂、路边，西北各地常见，属于优良牧草，放牧家畜特别是绵羊和山羊最为喜食。

91-1　二裂委陵菜　　　91-2　二裂委陵菜叶

A perennial herb, commonly found in wastelands, roadsides and field-sides, distributed in provinces of in northeast and northwest China as well as Sichuan province. The feeding value is high. Cattle, sheep and goats like to graze it any time.

92　多茎委陵菜

学名　*Potentilla multicaulis*
英名　Manystalk Cinquefoil

多年生草本。根发达，木质化，基部常有残余的托叶。茎多数，斜升或呈向内包弧形上升，密被柔毛。奇数羽状复叶，基生叶小叶数多于茎生叶。小叶矩圆形，羽状深裂，裂片条状。叶面深绿色，散生柔毛；叶背密生灰白色的绢毛。叶柄具长柔毛，托叶膜质。伞房状聚伞花序，花稀疏，花梗密生柔毛。花瓣5，黄色，宽倒卵形。瘦果椭圆状肾形，褐色，包被在宿存的花萼之中。多以伴生种出现于干草原、草甸草原及半荒漠草原。植株矮小，草质柔软，纤维含量低，适口性好，家畜喜食。

A perennial herb, it is commonly found in grasslands, tolerant to heavy grazing. It is good forage for horses, cattle, sheep and goats. It is distributed in north, southwest, northwest and northeast China.

92-1　多茎委陵菜　　　92-2　多茎委陵菜叶

93　鹅绒委陵菜

学名　*Potentilla anserine* L.
英名　Silverweed Cinquefoil

别名蕨麻、仙人果。多年生匍匐草本。根肥大，富含淀粉。纤细的枝条沿地表生长，呈匍匐状，叶丛直立、斜升或紧贴地面；奇数羽状复叶，小叶片卵圆形或披针状卵圆形，无柄，边缘具浅裂或整齐的深锯齿，叶背被白色绒毛，花单生于葶顶，花瓣5，卵圆形，鲜黄色。果瘦小，椭圆形，褐色。生长于高寒冷凉的环境，分布在青藏高原和毗临的黄土高原南缘。优良牧草，家畜喜食。

It is a prostrate perennial herb that spreads by crown buds and by seeds. It grows on meadows, swamp meadows, riverbeds and saline meadows and is distributed in north, northwest, northeast and southwest China. Cattle and sheep graze a little when fresh.

93-1　鹅绒委陵菜

93-2　鹅绒委陵菜花序

94　西山委陵菜

学名　*Potentilla sischanensis* Bunge
英名　Sishan Cinquefoil

多年生草本，种子繁殖。株高8~20厘米，全株除叶表面和花瓣外，几乎全被薄或厚的白色毡毛。根圆柱形，粗壮，黑褐色，根颈部包被残存的老叶柄。茎多数，丛生，直立或斜升。奇数羽状复叶，多基生，具长柄；小叶羽状深裂，无柄，顶生3小叶较大，裂片多，两侧裂片少或不裂；茎生叶不发达，无柄。聚伞花序，花排列稀疏。萼片卵状披针形，花瓣黄色；花柱近顶生，基部膨大。瘦果肾状卵形，多皱纹，无毛。主要分布在黄土高原及周边的内蒙古、河西走廊的石质地带或山地灌丛林缘。宜放牧，春、秋、冬三季各种家畜均喜食。

94-1　西山委陵菜

94-2　西山委陵菜根

A perennial herbaceous plant, commonly found on rocky steppes and grows among hassocks along hill-slopes. It is distributed in northern China and in Loess plateau. It is an excellent forage, various domestic animals will graze it.

95 龙牙草

学名　*Agrimonia pilosa* Ledeb
英名　Hairyvein Agrimonia

别名仙鹤草。多年生草本。根状茎褐色，横走。不分枝或上部分枝，有开展的长柔毛和短柔毛。不整齐的羽状复叶，具小叶，小叶间夹有小裂片，小叶倒卵形或椭圆形，先端尖，基部楔形，边缘有粗圆锯齿形，两面具长柔毛和腺点。托叶卵形，有齿，总状花序顶生，花黄色。生于路旁、林缘、河边及山坡草地，适口性中等，青草期牛乐食，羊、马、驴采食较少。

95　龙牙草

A perennial herbaceous plant, spread by crown buds and by seeds. It is commonly found in roadsides, forest fringes, river sides and mountain grassy slopes. It is distributed throughout the country. The feeding value is moderate and cattle find is palatable while horses and sheep have limited grazing.

96 水杨梅

学名　*Geum aleppicum* L.
英名　Aleppo avens

多年生草本。主根短，有多数须根。茎直立，上部分枝，羽状复叶，叶质薄。基生叶丛生，3～5 深裂或全裂，顶端一裂片最大。裂片近圆形、菱形或倒卵形，先端圆或尖，基部楔形，边缘常有不规则的深裂、浅裂或锐齿。侧生小叶 2～3 对，无柄。幼苗时叶面被毛后脱落，叶背毛随生长逐渐变少。茎生叶先端尖，边缘具不整齐的齿，托叶大，边缘具齿。花单生于枝端，花梗长，萼片披针形。花瓣黄色，倒卵形或近圆形。瘦果多数，长圆形，密被黄褐色毛，花柱宿存，先端呈钩状。鲜草家畜微嗜，干草采食较好。

A perennial herb, it is distributed in north, northwest and northeast China. The livestock animals do not graze.

96　水杨梅

97 地榆

学名 *Sanguisorba officinalis* L.
英名 Garden Burnet

别名山枣子、黄瓜香。多年生草本。根粗壮,茎直立,上部多分枝。奇数羽状复叶,小叶 5~15,矩圆状卵形至椭圆形,先端尖或钝,基部近楔形,边缘有圆锯齿。花小而密集,形成顶生圆柱形的穗状花序,每花有 2 苞片,披针形。萼片 4,紫红色。无花瓣;雄蕊 4,花柱短于雄蕊。瘦果宽卵形或椭圆形,棕褐色,有纵棱。主要分布在"三北"海拔较高的地区。适口性差,青鲜时家畜不喜食,其叶片和嫩枝调制成干草后,牛、绵羊、驴皆食。

A perennial herb reproduced from seeds.It grows in wastelands, grasslands and the edge of forest, distributed in south of the Loess Plateau. The feeding value is low.Cattle, sheep and goats would graze in winter, but not in summer.

97-2　地榆花序　　　97-1　地榆

98 绒毛绣线菊

学名 *Spiraea dasyantha* Bunge
英名 Hairyflower Spiraea

别名土庄、毛花绣线菊。灌木。高 1~2 米,小枝幼嫩时密被绒毛,后脱落,灰褐色。叶片菱状卵形,先端钝圆或锐尖,基部楔形,叶尖部边缘有深刻锯齿或裂片,上面深绿色,疏生短柔毛,下面被灰白色绒毛,羽状叶脉明显。叶柄短,密被绒毛。伞形花序具总梗,被灰白色绒毛。有花 10~20 朵,花白色。萼外密被白色绒毛,裂片三角形或卵状三角形。花瓣宽,近圆形,先端微凹。蓇葖果,开张,被绒毛。主要分布在我国陕西、甘肃、山西、江西、湖北等省,属于中旱生植物,生长在向阳山坡,常能形成建群种。春季返青后的嫩枝叶家畜喜食,尤其是山羊和绵羊。

A drought-tolerant shrub, growing on mountain forest, among shrubs and grasslands.It is distributed over all of northern China, including the Loess Plateau. Its tender leaves can be used as fodder for goats in the early spring.

98　绒毛绣线菊

99　珍珠梅

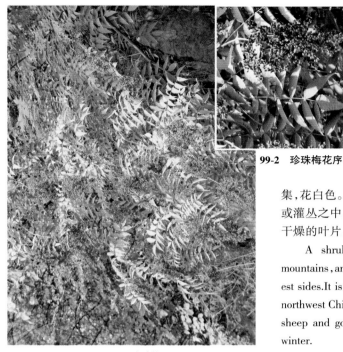

99-2　珍珠梅花序

99-1　珍珠梅

学名　*Sorbaria sorbifolia* L.
英名　*Kirilow falsepiraca*

落叶灌木，多丛生。茎中木质疏松，茎枝稍屈曲，新枝皮灰黄色，老枝黄褐色。奇数羽状复叶，小叶十余片，无柄，披针形至卵状披针形，先端锐尖，具锯齿，叶面无毛或近无毛。夏季开花，圆锥花序，蓬松硕大，小花多而密集，花白色。生于海拔 2 000 米以上的林下或灌丛之中。青鲜时家畜不喜食，家畜采食干燥的叶片。

A shrub,spread on shady side of the mountains,among thickets,grasslands and forest sides.It is distributed in north,northeast and northwest China.The feeding value is low.Cattle, sheep and goats like to graze its dry leaves in winter.

100　悬钩子

学名　*Rubus saxatilis* L.
英名　Raspberry

多年生攀缘灌木。茎密生紫红色的腺毛和短弯皮刺，棘手。小枝幼时常有黄色绵毛，随着生长逐渐脱落。掌状复叶，小叶 5 枚，长椭圆形或倒卵状。

圆锥花序顶生，下部有总状花序腋生，密生黄绒毛，花白色。果实着生于扁平或隆起的花托上，每个心皮有 2 胚珠，聚合小核果，球状，黑褐色。多生于山坡草地、地埂崖边以及沟壑底部的红土泻溜坡。因多刺毛而影响适口性，放牧家畜偶尔采食嫩叶，干燥凋落的叶片在冬春季节绵羊和山羊喜食。

100-2　悬钩子果　　　　100-1　悬钩子

It is a shrub and grows in gravel slopes and hilly valleys,distributed in south,southeast and central Loess Plateau.The feeding value is low.Sheep and goats would graze dry leaves in the winter.

101 扁核木

学名 *Prinsepia uniflora* Batal.

别名山桃、马茹。落叶灌木。枝灰褐色。具片髓；腋生枝刺青灰色，长5～10毫米，无毛，较直。托叶小，宿存；有时近无柄；叶片线状长圆形至椭圆状披针形，基部楔形，先端钝有短尖头，稀为渐尖，全缘，叶缘毛，稀具浅锯齿，两面无毛。花单生或2～3朵簇生于叶腋；萼筒浅杯状，萼裂片三角状卵形，果期反折；花瓣倒卵形，先端圆，白色；果近球形，暗紫红色，被蜡粉；果核两侧扁，表面有刻痕。生长于山坡、黄土崖边或川河间沙丘上。山羊、绵羊均喜食其落叶，果实人可以食用。

101-1 扁核木

101-2 扁核木果实

A shrub reproduced from seeds and by crown buds.It occurs on nearby villages, hill-slopes, precipice edges and field-sides.It is distributed in all over the Loess plateau.Goats will graze tender branches and leaves, other livestock animals can use its dry leaves in winter

102 鲜卑花

学名 *Sibiraea laevigata* (Linn.) Maxim.

102-1 鲜卑花　　102-2 鲜卑花花序

灌木。高约1.5米。小枝粗壮，圆柱形，光滑无毛，幼时紫红色，老时黑褐色；冬芽卵形，先端急尖，外被紫褐色鳞片。叶在当年生枝条多互生，在老枝上丛生，叶片线状披针形、宽披针形或长圆倒披针形，先端急尖或突尖，稀圆钝，基部渐狭，全缘，上下两面无毛，有明显中脉及4～5对侧脉；叶柄不显，无托叶。顶生穗状圆锥花序，苞片披针形，花瓣倒卵形，花盘环状，肥厚，蓇葖果5，并立，具直立稀开展的宿萼。生于高山、溪边或灌丛中，分布于青海（海晏、西宁）、甘肃（岷县、西固）、西藏（索县）等地。

A shrub.It is commonly found in water sides, damp grasslands and alpine meadows, and distributed in parts of Gansu, Qinghai and Xizang.

103　金露梅

学名　*Potentilla fruticosa* L.
英名　Bush Cinquefoil

别名木本委陵菜、药王茶。落叶灌木。多分枝,老皮纵向剥落。小枝红褐色,幼时被长柔毛。羽状复叶,通常有小叶5,稀3,上面1对小叶基部下延与叶轴合生;叶柄被绢毛或疏柔毛;小叶长圆形、倒卵状长圆形、卵状披针形,先端锐尖,基部楔形,全缘;托叶膜质,披针形。单花数朵生于枝端呈伞房状,萼片卵形,花瓣黄色;瘦果卵形,褐棕色。适应性广,我国南北均有分布。枝叶柔软,春季马、羊喜食,牛也采食,在优良牧草分布较少的地方,仍不失为有价值的饲用植物。

A shrub, commonly found on alpine grasslands, hill-slopes and among other shrubs. It is distributed in north and northeast China as well as Gansu, Inner Mongolia and Xizang. Goats and yak will graze tender branches and fresh leaves in spring, sheep will graze it in autumn.

103　金露梅

104　银露梅

学名　*Potentilla glabra* Lodd
英名　Glabrous Cinquefoil

灌木,高0.3~2米。奇数羽状复叶,小叶片椭圆形、倒卵状椭圆形或卵状椭圆形,顶端圆钝或急尖,基部楔形或圆形,边缘平坦或微向下反卷。顶生单花或数朵并生,花梗细长,被疏柔毛,花瓣白色,倒卵形。瘦果表面被毛。

生于山坡草地、河谷岩石缝、林缘灌丛及疏林中。分布在华北、陕西、甘肃、内蒙等地。银露梅枝叶富含营养,质地柔软,适口性好,春季羊、马喜食,牛乐食。

104　银露梅

A shrub, it is commonly found on alpine grasslands, hill-slopes and among shrubs or in the cracks between rocks around gullies. It is distributed in Qinghai-Xizang plateau and close to Loess plateau. The palatability is good and sheep and goats will graze it.

十八 豆 科

105　陇东苜蓿

当地习称庆阳苜蓿,国家定名为陇东苜蓿。适宜于黄土高原种植,根系十分发达,5年生主根可达5米以上。株高1米左右,半直立,单株分枝多,茎细而密;叶片小而厚;花序紧凑;荚果暗褐色,螺旋形,2~3圈;种子肾形,黄色。抗旱、抗寒性强。一般旱地每667米^{2}年产鲜草2 000~4 000千克。草地持久性强,长寿。主要分布在甘肃省的庆阳、平凉及周边,已有2 000年左右的栽培历史。

105　陇东苜蓿

It is a very good local variety suitable for seeding on sands and clay soil.It can resist cold,drought and alkali soil conditions but can not tolerate water-logging.It is distributed Qingyang and Pingliang and over the Loess plateau in Gansu province.The yield is high with long growing season.Palatability is good and nutritive value is high.Various domestic animals like to graze it.

106　紫花苜蓿

学名　*Medicago sativa* L.
英名　Alfalfa,Lucerne

别名紫苜蓿。多年生草本。主根圆锥形,侧根着生根瘤。根颈上丛生茎芽,由茎芽发育成茎。茎直立而光滑,圆形。第一片真叶为单叶,尔后所生皆为三出羽状复叶。小叶倒卵状长椭圆形,托叶较大,近于披针形。花序由叶腋中生出花梗,簇生总状花序,有短柄。花冠蝶形,花色以紫为主,兼有深紫、粉红、白、黄等。荚果螺旋形,有网纹,幼嫩时绿色,随着种子的成熟逐渐变成黄褐色、黑色。荚果含种子2~8粒,肾形,有光泽。紫花苜蓿被誉为"牧草之王",产草量高,营养含量丰富,饲用价值高,适口性好,马、牛、羊、猪、兔等各种家畜均喜食。

106-1　紫花苜蓿　　106-3　紫花苜蓿根

It has been cultivated for about 2 000 years in the Loess Plateau.The nutritive value is quite high and palatability is good so various domestic animals like to graze it.It is one of the best forages in the world.

106-2　紫花苜蓿花序

107 白花苜蓿

"白花苜蓿"是庆阳黄土高原试验站从紫花苜蓿中筛选出来的变异种,通过十余年的种植观察,与当地主栽的庆阳苜蓿相比,它具有前期生长快、叶片大、植株较高,第一茬产量高等特点。喜水肥,宜稀植;寿命和高产期的维持时间短。对牛、羊、猪、兔等家畜的适口性无明显差异。其特点非常适合黄土高原中南部草田轮作。

This is a variety of alfalfa (lucerne).It grows quickly in spring and early summer.The feeding value is quite high.All animals like to graze it all year round,fresh or dry.

107-1 白花苜蓿

107-2 白花苜蓿花序

108 黄花苜蓿

学名　*Medicago falcata* L.
英名　Sickle Alfalfa

别名野苜蓿、镰荚苜蓿。多年生草本。根粗壮,茎斜升或平卧,长 30~100 厘米。多分枝。三出复叶倒披针形、倒卵形或长圆状倒卵形,边缘上部有锯齿。总状花序密集成头状,腋生,花黄色,蝶形。荚果稍扁,镰刀形,稀近直立,被伏毛,内含种子 2~4 粒。喜湿润而肥沃的砂壤土,耐寒、耐风沙与干旱。多见于平原、河滩、沟谷及丘陵等低湿生境中,很少进入林缘。广布我国"三北"地区及内蒙古自治区。优良的饲用牧草,青鲜时,马、牛、羊最喜食,有催乳和促进幼畜生长发育的功效。

A perennial legume, reproduced from seeds and by rhizome.It is suited to fertile and wet sandy soils.It is commonly distributed in north, northwest and northeast China.The feeding value is high and the plant is palatable for most animals.

108-1 黄花苜蓿　　　　　　108-2 黄花苜蓿花序

109　圆形苜蓿

学名　*Medicago orbicularis* L.

别名圆盘荚苜蓿。一年生草本。株高10～50厘米，茎基部分枝，细弱，多平卧，稀直立，通常无毛。三出羽状复叶，小叶片倒卵状楔形，或倒三角形，边缘具小齿，上部边缘具齿明显，无毛或有极少的腺毛。总状花序腋生，由1～3花组成，花柄长，细弱，下垂。花萼钟状，花冠橙黄色，荚果圆盘状，由2～5螺旋盘曲而成，扁平，脊上具膜质边，果皮具疏松的网状脉，无毛或具腺毛。种子多数，卵形，黄褐色或灰褐色。多生于山地灌丛、荒漠沙地，有野生种也有栽培种。各种家畜的优质饲草。也可作为冬小麦收获后的轮作复种品种。

109-1　圆形苜蓿

109-2　圆形苜蓿花序

An annual legume reproduced from seeds. It is often used as multiple-purpose plants in some areas of the Loess Plateau. The feeding value is quite high. It is palatable for all livestock. It is a good forage for young animals in winter.

110　蒺藜状苜蓿

学名　*Medicago truncatula*

一年生草本。原产热带、亚热带地区，喜湿热环境。根系发达，多密集于耕作层。植株匍匐生长，枝条圆柱形，略带紫红色，疏被柔毛。分枝较短，分枝量较少，为放牧型牧草。在庆阳黄土高原每667米^{2}产鲜草1 000～2 000千克。生长期150天左右，也可在油菜或冬小麦收获后复种。三出复叶，具柄；小叶近菱形，两面被绵毛，中间有一黑斑，叶缘密具长尖齿；托叶轮廓三角形，深裂，膜质。螺旋荚果紧密成球形，外生刺状突起，形如蒺藜。成熟后荚果易脱落，种子圆肾形，淡黄色。草质良好，各种家畜喜食。

110-1　蒺藜状苜蓿

110-2　蒺藜状苜蓿花序

An annual legume, reproduces from seeds. It is usually cultivated in warm and wet region of the world. The texture of the grass is soft, palatability is good and nutritive value is high. Most livestock animals like to graze anytime.

111　红豆草

111-2　红豆草花序

学名　*Onobrychis viciaefolia* Scop.
英名　Common Sainfoin

别名驴食豆。多年生草本。主根短而粗壮，侧根发达，根瘤成串且大。茎圆形直立或半倒伏，从根茎和叶腋处分枝。奇数羽状复叶，小叶片长椭圆形至线形，着生于总叶柄的两侧。穗呈总状花序，每序有小花40～60朵，呈四列排在花轴上。花冠以紫红色为多，亦有粉红、紫色和白色者。荚果黄褐色，外布凸起的网状皱纹，半圆形或卵形，边缘有锯齿状薄片。每荚含一粒种子，种子肾形，绿褐色，表面光滑。较耐干旱。该草营养丰富，早春生长较苜蓿快，秋季适口性明显增加。

111-1　红豆草

A perennial legume, it is suited to arid regions of the Loess Plateau. It has higher nutritive value at the flowering stage. Animals will graze in early spring and late autumn. No bloating problem when grazed fresh.

112　红三叶

学名　*Trifolium Pratense* L.
英名　Red Clover

别名红车轴草。多年生草本。茎直立或斜向上，中空，有毛或无毛。主根入土深，侧根发达，根瘤数量多。叶3片，聚生于茎柄顶端，全缘；边有绒毛，叶柄长，叶片上有"V"形斑；小叶卵形或椭圆形；托叶大，先端尖锐，膜质，有紫色脉纹。头形总状花序，着生于茎枝顶端或叶腋处；花小，每序50～100朵。花冠红色或淡红色，萼膜质有毛，呈钟形。荚果小，卵形，含一粒种子；种子肾形，黄色。原产于小亚细亚及南欧，我国新疆、云南、贵州、湖北、四川等地均有分布。以鄂西栽培时间最长，云、贵面积最大。系优良牧草之一，家畜喜食。

112-1　红三叶

112-2　红三叶花序

It is a perennial legume, cultivated over all parts of the country. It is a good forage in the southern Loess Plateau. The feeding value is high and all livestock graze this plant.

113 白三叶

学名　*Trifolium repens* L.
英名　White Clover

别名白车轴草。多年生草本。茎枝匍匐贴地面,茎节上生不定根,并形成新的株丛,繁殖力和竞争力极强。主根入土较浅,侧根发达。茎细长柔软,基部分枝多。光滑无毛,有实心白髓。三出复叶,叶柄细长。小叶倒卵形或倒心形,叶片中央有"V"形白斑或无。叶缘有锯齿,托叶小,膜质。头形总状花序,生于叶腋间,花柄细长,花

113-3　白三叶

小而多,白色或粉红色,花冠不脱落。荚果细而长,每荚有种子3~4粒,种子心脏形。原产欧洲,我国云、贵、川、浙、湘、鄂以及东北吉林、西北新疆均有野生种。叶片多,营养丰富,家畜喜食,适宜于草食家畜放牧。

113-1　白三叶根茎

113-2　白三叶花序

　　It is a prostrate perennial legume that spreads by crown buds and by seeds. It is distributed in east, south and central Loess Plateau as well as southern Yellow River. The feeding value is high with domestic animals grazing after a frost in the autumn.

114 沙打旺

学名　*Astragalus adsurgens* Pall
英名　Erect Milkvetch

别名直立黄芪。多年生草本。主根粗壮,发达,入土深,侧根繁盛,上面着生大量根瘤。茎中空,直立或斜生。分枝多,主茎不明显,全株被丁字形茸毛。奇数羽状复叶,卵状椭圆形。总状花序,多数腋生,亦有顶生者;花冠蓝色、紫色或蓝紫色,蝶形。荚矩形或圆筒形,2室,内有种子4~10粒,黑褐色。原产于黄河故道地区,即山东、河南、河北和苏北一带,是我国特有的草种。因其耐旱耐寒的性质,西北地区广有栽培。青鲜时适口性较差,调制青干草或加工草粉后是家畜冬季的优质饲草。

　　A perennial legume reproduced from seeds. It is usually cultivated in arid regions of the Loess Plateau. It has high nutritive value and it is better fed as hay or powder.

114-1　沙打旺

114-2　沙打旺花序

115　黄花草木樨

115-1　黄花草木樨

115-2　黄花草木樨花序

学名　*Melilotus officinalis* Desr.
英名　Yellow Sweetclover

别名马首蓿、野首蓿、金花草。一年生或越年生草本。根系粗壮发达，入土深，根瘤丰富。植株高大，直立，可达2米。茎圆中空，光滑或稍有毛，白花种植株比黄花种更高大。三出复叶，小叶椭圆形或倒卵形，边缘有锯齿；托叶披针形，先端尖锐。总状花序，具短柄；旗瓣与翼瓣等长。荚倒卵形，光滑无毛，有网状皱纹，内含一粒种子（偶有含2粒种子者），种子长圆形或肾形，略扁平，棕黄色，有特殊异味。可放牧、青饲、调制干草或青贮。因香豆素含量较高，所以适口性差，调制成干草粉可以极大地提高饲喂效果。

A biennial or annual legume reproduced from seeds. It grows well on low moist habitats such as wetlands and gullies. Feeding value is high and it is palatable for grazing livestock when it is immature, but it becomes coarse and woody and animals avoid it after flowering.

116　白花草木樨

学名　*Melilotus albus* Desr.
英名　White Sweetclover

越年生或一年生草本。茎直立，圆柱形，中空，有淡清香味。羽状三出复叶，小叶椭圆形或长圆形，先端钝，基部楔形，边缘有疏锯齿。托叶较小，锥形或条状披针形。总状花序腋生，花小，白色，花萼钟状。花蝶形，旗瓣长于翼瓣。小荚果，椭圆形。种子肾形，黄色或褐黄色。黄土高原西北部有野生种，但近年来栽培种较多，耐瘠薄，产量高。属于优良牧草，开花时香豆素含量高，影响适口性和利用率。特别值得注意的是发霉或腐败后香豆素会转变为抗凝血素，可引起家畜出血死亡。

A biennial or annual legume reproduced from seeds. Its yield is higher than the Yellow Sweetclover, and feeding value is high. Cattle, sheep and pig.

116-1　白花草木樨

116-2　白花草木樨花

117　多变小冠花

学名　*Coronilla varia* L.
英名　Crown Vetch

别名小冠花。多年生草本植物。根系十分发达，质嫩，色黄白，生长 2 年，主根入土深达 1 米以上，根幅可达 2 米以上。侧根繁多，横向走串，主侧根均可长出不定芽，进行无性繁殖，根瘤丰富。根茎粗大，分生能力和再生能力极强。茎中空有棱，柔软。叶为奇数羽状复叶，互生，小叶长卵形或倒卵圆形。伞形花序，腋生，由十多朵小花组成，花色粉红。小花呈环状紧密排列在花梗顶端。荚果细长，上有节，成熟后易从节处断裂。每节有一粒种子，种子细长，红褐色。原产南欧和中东地中海，该草营养物质含量丰富，但含有毒物质——β 硝基丙酸，单胃家畜不能采食过多。若调制成青干草饲喂反刍家畜，其瘤胃微生物可以分解毒素。

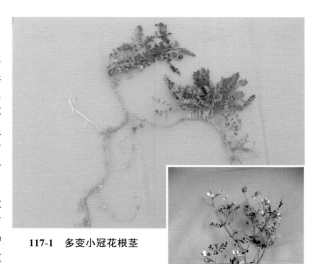

117-1　多变小冠花根茎

117-2　多变小冠花

It is a prostrate perennial legume that spreads by crown buds and by seeds. It is usually cultivated on the slopes and gullies of the Loess Plateau. When it is fresh, sheep and goats graze a little, but will eat as hay in winter.

118　百 脉 根

学名　*Lotus corniculatus* L.
英名　Bird's Foot Trefoil

118-2　百脉根荚果　　**118-3　百脉根花序**

118-1　百脉根

别名五叶草、鸟足豆、牛角花。多年生草本植物。直根系，主根发达，侧根繁多。茎枝丛生，表面光滑无毛，匍匐或直立生长。掌状三出复叶，小叶倒卵形或卵形，短叶柄。二托叶较大，与小叶相似，有时被误认为是 5 片叶子。蝶形花，黄色，旗瓣有明显紫红色脉纹。伞形花序，有小花 4~8 朵，位于长花梗顶端。荚长而圆，角状，似鸟足，成熟后为褐色。每荚有种子 10~15 粒，种子小，黑褐色，有光泽。原产欧亚大陆的温暖地带，我国华南、华北、西南均有种植。营养价值高，适口性好，各种家畜均喜食。

A perennial semi-prostrate legume reproduced from seeds. It grows well under the full sun, suited to dry-lands. It is distributed in Gansu, Shaanxi and southwestern China. The texture of the stem is soft, with a large amount of leaves. Various domestic animals like it.

119 紫云英

学名 *Astragalus sinicus* L.
英名 China Milkvetch

119-1 紫云英

119-2 紫云英花序

一年生草本植物。根系发达,主根肥大,侧根多。茎中空,直立或匍匐。奇数羽状复叶,小叶圆形或倒卵形,全缘,先端圆或微凹,中脉明显,叶面光滑或疏生短茸毛,深绿色,叶背有绒毛,色浅淡。伞形花序,花梗细长,自叶腋生出,淡红色或紫红色。荚果条状长圆形,有隆起的网脉,稍弯,先端有喙,成熟时黑褐色,内含种子5~10粒,种子肾形,光滑,黄绿色。茎叶柔嫩,适口性好,马、牛、羊、兔、猪、禽均喜食,也可作为鱼类的饲料。

A biennial or annual legume reproduced from seeds. It is distributed around the Yangtze River valleys, as well as southern Loess Plateau. The feeding value is high and the plant is palatable for most animals, particularly pigs.

120 箭筈豌豆

学名 *Vicia sativa* L.
英名 Common Vetch, Fodder Vetch

别名春巢菜、大巢菜、救荒野豌豆。一年生草本。主根明显,有根瘤。茎柔嫩有条棱,有细软毛或无,匍匐向上或半攀缘状。偶数羽状复叶,呈矩形或倒卵圆形,先端凹入,中央有突尖。叶轴顶端有分枝的卷须,缠于它物上。托叶半箭形或戟形,有1~3枚披针形裂齿。蝶形花1~2朵,腋生,紫红、粉红或白色,花梗极短或无。荚果扁,成熟时为黄色或褐色,含种子5~12粒,种子扁圆或钝圆,有黄、白、灰、粉红、黑等多种颜色。原产南欧、西亚,喜温凉、抗严寒,目前全世界广有种植。从20世纪50年代开始把它作为主要的复种、轮作牧草,播种面积仅次于紫花苜蓿。属于饲草、饲料兼用作物,产量高,营养丰富,适口性好,各种家畜都喜食。

An annual semi-prostrate legume reproduced from seeds. It is an excellent forage with high feeding value, distributed throughout the country. It can be used as hay for young animals over winter.

120-1 箭筈豌豆

120-2 箭筈豌豆花序

121 毛苕子

学名 *Vicia villosa* Roth
英名 Villose Vetch, Hairy Vetch

别名毛野豌豆、冬箭筈豌豆、冬巢菜。一年生或多年生草本。子叶不出土,茎叶由胚芽发育而成。根系发达,主根明显,侧根分枝多;根瘤较多,呈扇形、姜形、或鸡冠形。茎四棱中空,匍匐蔓生,基部有 3 ~ 6 个分枝节,全株密生银灰色白绒毛。偶数羽状复叶,复叶狭长,顶端有卷须 3 ~ 5 个。总状花序,每个花梗有小花 10 ~ 30 朵,聚生于花梗上部的一侧,花柱上部四周有长柔毛,花冠紫蓝色。花萼斜钟状,萼齿较长,有茸毛。荚果短矩形,两侧稍扁,具不明显网脉,色淡黄,光滑,易爆裂。每荚含种子 2 ~ 8 粒。种子圆形,黑褐色,种脐色略淡。喜生于水热条件较好的地方,普遍用于夏季复种、压青倒茬或为幼畜准备越冬饲草。

121-1 毛苕子

121-2 毛苕子花序

A perennial or annual creeping legume reproduced from seeds. It is mainly distributed in southern China and can find it in some regions of the Loess Plateau. The texture of the stem is soft, palatability is good and feeding value is high with domestic animals.

122 美国香豌豆

学名 *Lathyrus odoratus* L.

英名 American Vetchling

多年生香豌豆系,1998 ~ 2000 年从美国内布拉斯加(Nebraska)引入的优良品种。主根粗壮,圆柱形,入土深达 2 米以上;根颈位于土表下 5 ~ 10 厘米;具团聚状根瘤,多着生于侧根上。蔓生,高约 140 厘米,匍匐茎,扁平状,具卷须,可缠绕攀缘。叶互生,呈椭圆状披针形,全缘,具叶柄;顶生小叶退化成卷须状,叶基渐狭,叶尖凸尖。无限总状花序腋生,具小花 3 ~ 7 朵。花瓣离生,5 片,花色有蓝、紫、红、粉、白等多种。上位子房同位花,多胚珠,子房一室,边缘胎座,自花授粉。荚果长矩形,表面光滑,有脉纹,幼嫩时为绿色,成熟后为黄褐色,极易开裂。每荚含种子 4 ~ 18 粒,多数为 13 ~ 15 粒,种子绿灰色。庆阳黄土高原试种结果表明:喜水肥,生长快,每 667 米^{2}产鲜草 2 000 千克左右,适口性中等,牛、羊、马喜食。花色多样,鲜艳,花期长,具有较高的观赏价值。

122-2 美国香豌豆花序

122-1 美国香豌豆

A perennial trailing legume reproduced from seeds. It grows quite well in Qingyang Research Station. The feeding value is high and it is one of the best forages in the word. The texture of the plant is soft, with a large amount of leaves. Various domestic animals like it.

123 山羊豆

123-3 山羊豆根系

123-1 山羊豆

123-2 山羊豆花

学名 *Galega officinalis* Linn.
英名 goat'rue

多年生丛生草本。地下根茎横走，发达，近地表节上易形成新的植株。茎直立，中空。叶互生，奇数羽状复叶，蜡质，光滑，小叶极多数，全缘；托叶箭头状或叶状。花排成顶生或腋生的总状花序；苞片小，刚毛状，常宿存；萼齿近相等；花冠白色、紫色或蓝色；子房无柄，有胚珠多数；荚果线形，圆柱状。种子肾形，黄色或黄褐色。庆阳黄土高原试验站于1997年从白俄罗斯引进栽培，喜水肥，不耐高温，易形成坚硬的草结皮抑制了自身的生长。第一年每667米²产鲜草2 000千克，以后逐年迅速锐减。因茎秆易粗老，引起适口性下降，家畜较喜食。

A perennial legume reproduced from seeds and by rhizome.It is a good forage that grows at Qingyang Research Station.It originated from Russia.It has higher nutritive value in the flowering stage.Cattle and sheep like to graze all year around.

124 山黧豆

学名 *Lathyrus sativa*
英名 Grass Peavine

别名栽培山黧豆、马牙豆。一年生草本。根系发达，入土较深。茎多斜升，自下部分枝。偶数羽状复叶，具小叶一对，叶轴顶端具卷须。小叶条形或披针状条形，先端渐尖，基部渐狭。叶背脉突起明显，近平行。花腋生，1~2朵，花梗短，白色或蓝色等。花萼宽钟形，萼齿5，披针形。蝶形花冠，荚果长圆形，微扁，内含种子2~5粒。种子白色、灰色或褐色，为不规则的三角形、四棱形或楔状四棱形。主要分布在陕北和甘肃中部干旱地区。茎叶柔嫩，适口性好，营养丰富，家畜喜食。子实中含有β-草酰氨基丙氨酸，有毒，饲喂家畜时要用清水浸泡或蒸煮脱毒。

An annual legume,distributed over the Loess plateau.The straw is an excellent forage.Palatability is good and nutritive value is high.Various domestic animals like to graze.The seeds can be used as livestock concentrates.

124-1 山黧豆   124-2 山黧豆花序

125 花 棒

学名　*Hedysarum scoparium* Fisch
英名　Slenderbranch sweetvetch

别名细枝岩黄芪、花柴、花帽。多年生大灌木。高可达2米,茎和下部枝紫红色或黄褐色,常呈纤维状剥落。小枝绿色,多分枝。小叶多披针形、条状披针形,稀长圆形;上部枝小叶较少,较窄,或仅有叶轴而完全无小叶。托叶卵状披针形,结合,伏生毛早落。总状花序稀疏,花梗腋生;苞片小,三角状,密被柔毛;萼筒钟形或披针状钻形;花冠紫色。荚果2~4节,荚节凸胀,近球形。种子耳状。耐风蚀,喜沙埋,分布在甘肃、新疆、宁夏和内蒙古。营养价值高,适口性好,山羊、绵羊、骆驼喜食其嫩枝叶。

A tall shrub, commonly found in saline soils in desert region, and resistant to cold, drought and alkali conditions. It is distributed in Gansu, Ningxia and Inner Mongolia. Palatability is good and nutritional value is high. Camel, sheep and goats will graze leaves, fruits flowers and tender branches.

125-1　花棒　　　125-2　花棒花序

126 紫穗槐

学名　*Amorpha fruticosa* L.
英名　Falseindigo, Indigobush Amorpha

别名紫花槐、棉条。落叶灌木。丛生,直立,枝叶繁茂,皮暗灰色,平滑,小枝灰褐色,有突起的锈色皮孔,幼小时密被柔毛。侧芽很小,常2个叠生。叶互生,奇数羽状复叶,小叶11~25,卵形、椭圆形或狭椭圆形,全缘,叶内有透明油腺点。总状花序密集,顶生或在枝端腋生,花轴密生短柔毛。花萼钟形,常有油腺点。旗瓣蓝紫色。荚果短弯曲,棕褐色,密被瘤状腺点,不开裂,内含1粒种子。分布在我国包括黄土高原在内的北方,耐寒旱,抗风沙,枝叶含有较高的营养,晒干后是各种家畜的优质饲料。

A shrub, it is cultivated or volunteer plant, occuring in wastelands, hill-slopes and gullies of the Loess Plateau. With large amount of leaves, palatability is good and nutritive value is high. Various domestic animals like to graze.

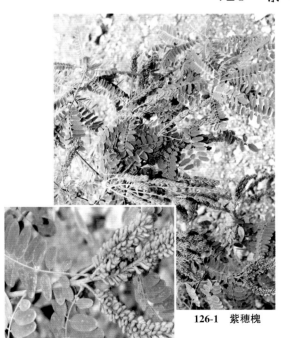

126-1　紫穗槐

126-2　紫穗槐花序

127 豌豆

学名 *Pisum sativum* Linn.
英名 Garden Pea

127-1 豌豆

一年生缠绕草本，高 90～180 厘米，茎圆柱形，中空而脆。全体无毛，被白霜。双数羽状复叶，小叶 2～6 片，长圆形至卵圆形，全缘；托叶叶状，卵形，通常大于小叶，基部耳状包柄。花单生或 1～3 朵排列成总状而腋生；花冠白色或紫红色；花柱扁，内侧有须毛。荚果长椭圆形；种子圆形，青绿色，干后变为黄色。在我国栽培历史悠久，分布普遍，以甘、陕、宁、青、晋、冀较多。营养价值高的传统饲料作物，可青饲、青贮、调制干草或干草粉，各种家畜均喜食。

An annual legume, reproduces from seeds. It has been cultivated for long time in the Loess plateau. The feeding value is high, the straw is soft and palatability is good, the seed is traditional concentrate for all livestock animals.

127-2 豌豆花序

128 扁豆

学名 *Lablab purpureus* (Linn.) Sweet
英名 Hyacinth Bean

别名蛾眉豆、鹊豆、沿篱豆。一年生或越年生草本。根系发达，茎蔓生有分枝，多直立丛生，光滑或有毛。三出复叶，具柄，披针形，顶生小叶菱状广卵形，顶端短尖或渐尖，基部宽楔形或近截形，两面沿叶脉处有白色短柔毛。总状花序腋生，花 2～4 朵丛生于花序轴节上；花冠白色或紫红色，旗瓣基部两侧有 2 附属体；子房有绢毛，基部有腺体，花柱近顶端有白色髯毛。荚果扁，镰刀形或半椭圆形。种子 3～5 粒，扁，长圆形，白色或紫黑色。短日照植物，喜温，怕寒，适宜肥沃的沙壤土。我国南北均有分布，但多为零星种植。主要的饲料和粮食作物，秸秆含有丰富的蛋白质，是各种家畜的优质饲料。

128-1 扁豆

A small annual legume, reproduces from seeds. It is suited to hill slopes in Loess plateau. The feeding value is high and the ruminant animals will graze. In addition, the seed is edible.

128-2 扁豆荚

129 蚕豆

学名 *Vicia faba* Linn.
英名 Horsebean, Broadbean

别名佛豆、胡豆、马齿豆。一年生草本。茎直立,无毛,高30~180厘米。羽状复叶有小叶1~3对,顶端卷须,不发达而为针状;托叶大,半箭头状。花1至数朵,腋生在极短的总花梗上;花白色带红,有紫色的斑块。萼钟形,萼齿5,披针形。荚果大而肥厚。种子卵圆形,略扁。原产里海南部至非洲北部,张骞出使西域时引入我国,现西南各地栽培较多,甘肃、青海也大量种植。茎叶可作青饲料,猪喜食,马、牛采食,种子则是役畜的重要精饲料。

129-1 蚕豆

129-2 蚕豆花序

An annual legume, reproduces from seeds. It is suited to fertile and wet soils of cool region. It is distributed in southwest Loess plateau and southwestern China. The feeding value is high and the seed is traditional concentrate for more livestock animals.

130 莲山黄芪

学名 *Astragalus leansanicus* Ulbr
英名 Leansan Milkvetch

多年生草本,茎丛生,有角棱,被极疏的白色丁字毛。奇数羽状复叶互生,小叶卵状披针形或披针形,先端圆或尖,基部宽楔形,全缘,叶背疏被贴伏的白色粗毛。托叶合生,无毛。总状花序腋生,有花多朵,总花梗稍长于叶。花萼钟状,裂片5,有丁字毛,蝶形花冠淡红色或近白色,旗瓣稍长,,先端深凹。荚果棍棒状,渐尖,具短喙,疏被粗毛或无毛。种子肾形,中部凹陷,黄褐色。分布于黄土高原的陕、甘两省,为我国西部的特有品种。草食家畜喜食。

130-1 莲山黄芪

130-2 莲山黄芪花序

A perennial legume reproduced from seeds. It is commonly found in sandy areas of river beaches and wastelands. It is commonly found along fieldsides, near orchards and among hassocks along road-sides. It is distributed in the western part of China. It is a good forage, all livestock like it when fresh.

131　鸡峰山黄芪

学名　*Astragalus bhotanensis* Baker.
英名　Jifengshan Milkvetch

多年生草本。主根系发达。茎丛生,多分枝,斜升或匍匐,基部有短柔毛。奇数羽状复叶,互生;小叶3~9片,几无柄,叶片狭状披针形,先端稍尖,基部楔形,全缘,上面无毛,叶脉和叶背疏生白色丁字柔毛。腋生总状花序,花冠淡红色或白色。荚果长,圆柱形,被丁字毛,先端有喙。种子长圆形,棕黑色。中等牧草,各种家畜都喜食,春季羊最喜食其花序。

A perennial legume reproduced from seeds.It is commonly found in hill slopes and wastelands, and distributed in Loess plateau and the Yunnan-Guizhou plateau.The feeding value is high before flowering and various domestic animals will graze it.

131-1　鸡峰山黄芪

132　糙叶黄芪

学名　*Astragalus scaberrimus* Bge
英名　Coarseleaf Milkvetch

别名春黄芪、掐不齐。多年生草本。具有横走的地下根状茎,有地上茎或无地上茎;有地上茎也是很短且平卧于地面,似莲座状,叶密集于地表。通体被白色的丁字毛,奇数羽状复叶,有小叶7~15片。小叶椭圆形、矩圆形或椭圆状披针形,全缘。托叶由叶柄向下延长而成,宽大,占叶柄的1/3或1/2。总状花序由基部腋生,具蝶形花3~5朵,花冠白色、淡黄色或淡紫红色,长1~2厘米。花萼筒状,萼齿条状披针形。旗瓣椭圆形,先端微凹,中部以下渐狭,翼瓣和龙骨瓣较短,子房有短毛。荚果圆筒状,稍弯。是黄土高原干旱针茅草原的主要伴生种。春季开花时,山羊或绵羊喜食其花和嫩叶,夏秋季喜食其果实。由于植株矮小大家畜几乎无法采食。

A perennial herb,spread by crown buds and by seeds.It grows on the gravel of the Gobi desert and clay soils of desert regions.It is distributed in north,northwest and northeast China.The nutritive value is high.Goats, sheep and small animals like to graze when fresh or dry and its seeds.

132-1　糙叶黄芪

132-2　糙叶黄芪花

十八 豆 科

133 草木樨状黄芪

学名 *Astragalus melilotiode* Pall
英名 Sweetcloverlike MilkVetch

别名扫帚苗。多年生草本。主根粗壮而深。茎直立或稍斜升,多分枝,有棱,被短毛或无毛。奇数羽状复叶,小叶多数,长圆状楔形或线状长圆形。总状花序腋生,花序长,花多而疏生,花冠粉红色或白色。荚果近圆形,先端钝,具弯曲短喙。适口性良好,尤以春季幼嫩时家畜喜食,花后迅速粗老,适口性变差,可食率降低。

It is a perennial erect more branching legume, reproducing from seeds, occurs in field sides, hill areas and roadsides. It is distributed throughout the Loess Plateau. The feeding value is high. Cattle and sheep like to graze its tender branches and camels like to graze it year around.

133-2 草木樨状黄芪花序　　133-1 草木樨状黄芪

134 太白岩黄芪

学名 *Hedysarum taipeicum*
英名 Taibai Sweetvetch

灌木。茎直立,多有分枝,细长,表面土褐色至土红棕色,表皮易剥落,有纵纹及横向皮孔。奇数羽状复叶,小叶椭圆状披针形,全缘,具极短柄,中脉明显,先端钝圆,基部楔形。花序轴生于叶腋,长于总叶柄;小花单生于序轴,有短柄,花冠白色至淡黄绿色;萼齿较长,与筒部等长或更长;荚果串珠状,多为2~5,圆形或椭圆形,无毛。生于山坡、林下或路旁,分布在陕西及毗邻的甘肃南部。属于优良饲用植物,家畜喜食其嫩枝叶、花序和荚果。

A perennial legume, reproduces from seeds. It grows on mountain area of Qinling and southern Loess plateau. The feeding value is moderate and livestock animals will graze.

134-1 太白岩黄芪　　134-2 太白岩黄芪花序

135　野大豆

学名　*Glycine soja* Sieb.et Zucc.
英名　Wild Soybean

一年生蔓生草本。茎细长，缠绕。三出复叶，顶生小叶卵状披针形，两面生白色短柔毛，侧生小叶斜卵状披针形，有托叶。总状花序腋生，花小，花梗密生长硬毛，花萼钟状，萼齿5，花冠白色或紫红色。荚果小，矩形，密生黄色长硬毛，内含种子2~4粒。种子近椭圆形，两侧稍扁，粗糙，黑色。黄土高原农田、沟壑湿地中易见，品质优良，富含营养，草质柔软，适口性极好，各种家畜喜食。

135　野大豆

An annual creeping legume.It is commonly found in river beaches, canals, roadsides, orch- ards and under forestry.It is distributed in south-eastern Loess Plateau as well as southern China, such as Hebei, Shandong, Anhui, etc.The feeding value is high.Most domestic animals graze all year around.

136　野绿豆

学名　*Vigna radiatus* L.
英名　Willd Green gram

一年生草本。茎蔓生、半蔓生或直立，幼茎呈紫色或绿色，被淡褐色刚毛。分枝1~5，顶端有卷须。三出复叶互生，有较长叶柄，小叶阔卵形至菱状卵形，顶部尖，基部圆形、斜形或截形，全缘，少有三裂及缺刻者，两面被毛。托叶着生于复叶柄基部，盾状，有数条隆起的粗脉，边缘有毛。总状花序顶生或腋生，总花梗密被长毛。花黄色或黄绿色，10~25朵小花簇生于花梗上部。蝶形花，其中有一龙骨瓣有角。荚果黑色、褐色、或黄褐色，荚长4~12厘米，内含种子6~15粒。种子球形或圆柱形，通常为绿色，亦有黄色、棕褐色或青蓝色。黄土高原零星分布，耐干旱，多生于弃耕地、荒地、农田和果园。有大量的栽培种，营养丰富，全株均为家畜的优质饲料。

An annual legume reproduced from seeds.It grows on farmland and wasteland, distributed in south and central Loess Plateau.The texture of the plant is soft with large amount of leaves, palatability is good and nutritive value is high.

136　野绿豆

137　歪头菜

学名　*Vicia unijuga*
英名　Pair vetch, Two-leaved Vetch

别名对叶草。多年生草本。根状茎发达。茎直立，通常数茎丛生，有棱。双数羽状复叶，具 2 小叶，叶轴末端成刺状。托叶半边箭头形，小叶卵形至菱形，大小和形状变化大，先端尖，基部圆或楔形，全缘，叶脉明显。总状花序腋生或顶生，具花 15～25 朵，花冠蓝紫色，花萼斜钟形，花冠蝶形。荚果扁平，长圆形，内含种子 1～5 粒。主要分布于"三北"地区，喜阴湿，多生于林下灌丛，山沟谷地草坡。营养丰富，适口性好，再生力强，适宜放牧，马、牛最喜食，家兔、鹿、羊亦乐于采食。

137-2　歪头菜花序　　　137-1　歪头菜

A perennial herb reproduced from seeds. It occurs in gullies, slopes, grasslands and forest, distributed in northeast, southeast and southwest China as well as north China such as Shaanxi. It can tolerate to heavy grazing and is good forage for cattle, horses, sheep and goats.

138　甘　草

学名　*Glycyrrhiza uralensis* Fisch.
英名　Ural Licorice

多年生草本。根粗壮且十分发达，入土深，横走，长，圆柱形，有甜味。茎直立，多分枝，具白色短柔毛和腺体。羽状复叶互生，具长柄。小叶卵形或宽卵形，全缘，先端急尖或钝，基部圆形，两面均有短毛和腺体。总状花序腋生，花密集。蝶形花冠紫红色或蓝紫色。荚果条形，呈镰刀状或环状弯曲，密生刺毛状腺体。每荚有种子 4～8 粒，肾形，可用于繁殖。喜沙性稍干燥的钙质土壤，多生于半荒漠草原、荒地和农田，是著名的中药材，家畜喜食。

A perennial grass, it is spread by rhizome and by seeds. It grows in sunny drought prairies with calcareous soil, river banks with sandy soil, farmlands and roadsides. It is distributed in north and northwestern China. The feeding value is moderate. Camels graze before budding, and other livestock would graze it after it is dry. Sheep and goats like its pods.

138-1　甘草　　　138-2　甘草花序　　　138-3　甘草果荚

139 米口袋

学名 *Gueldenstaedtia multiflora* Bge.
英名 Rice bag

多年生草本。茎短缩，数条簇生于根颈上。羽状复叶，小叶椭圆形，先端钝圆或略凹，有小尖头，基部圆形，两面均有毛。托叶三角形，上有长柔毛。伞形花序有花 4～6 朵，花葶较长。花萼钟状，密生长柔毛，花萼上二齿较大，花冠紫红色，旗瓣卵形，长于翼瓣，龙骨瓣短。荚果圆筒状，有毛，1 室。种子肾形，黑褐色，有凹点。该草从东北到西北均有分布。牧草品质好，多种家畜喜食。

A perennial legume reproduces from seeds. It occurs in hill slopes, grasslands and orchards, distributed in provinces of north and northeast China as well as Gansu, Shanxi, etc. It is a good quality forage and palatable for sheep and goats.

139-1 米口袋花序

139-2 米口袋

140 狭叶米口袋

140 狭叶米口袋

学名 *Gueldenstaedtia stenophylla* L.
英名 Narrow-leaf Gueldenstaedtia

多年生草本。全株被稀疏灰白色长柔毛。茎短缩，数条簇生于根茎上。羽状复叶，小叶长椭圆形或条形，先端有短尖。托叶宽三角形或三角形，密生柔毛。伞形花序，有花 2～3 朵。花葶长，花萼钟状，有毛，花冠粉红色或近白色，荚果圆筒形，有毛，1 室。种子肾形，有凹点。分布在黄土高原南部，生于山坡、草地、路旁或农田，家畜喜食。

A perennial legume reproduced from seeds. It is commonly found in farmlands, hill slopes, grasslands and roadsides. It is harmful to dry-lands crops and nursery gardens. It is distributed over Loess Plateau as well as southern China, such as Henan, Jiangsu, Jiangxi. The feeding value is high and most livestock like it all year around.

141　少花米口袋

学名　*Gueldenstaedtia verna* (Georgi) Boriss.
英名　Few-flower Gueldenstaedtia

多年生草本。茎短缩，托叶狭三角形，基部合生，被白色长柔毛。叶柄具沟槽，花后伸长；小叶窄卵形至宽卵圆形，顶端小，有时为倒卵形，先端圆或微凹陷，具短尖，基部近圆形，全缘；两面被白色长柔毛或表面毛较少。花梗短，小苞片2，条状披针形。花冠紫红色。荚果圆柱形，种子表面有凹点。旱生植物，生于草原带的沙质草地、石质草地或草坡。属优等饲草，绵羊、山羊都采食。

141　少花米口袋

A perennial legume, reproduces from seeds. It exists in wastelands, orchards and hill slopes, distributed in north, northeast and northwest China as well as Gansu and Shanxi, etc. The feeding value is moderate, sheep and goats like to graze.

142　长叶铁扫帚

学名　*Lespedeza caraganae* Bge.
英名　Longleaf Bushcloyer

小灌木。根粗壮，淡褐色。茎自基部分枝，斜生或近直立，被白色短柔毛。三出复叶，互生，小叶条状长圆形或线状矩形，有短尖，两面疏生白柔毛。小叶柄极短或近无柄，托叶狭细。总状花序生于枝条中上部叶腋，总花梗劲直，短于叶。花萼杯状，有毛，蝶形花冠白色，基部淡红色。荚果卵圆形，渐狭，种子长圆状倒卵形，黑紫色。生于地埂、山坡。适口性较好，春季幼嫩时家畜喜食。

A semi-shrub reproduced from seeds. It is commonly found in mountain area, and among hassocks along roadsides. It is distributed in south and southeast Loess Plateau. The feeding value is high. Cattle, sheep and goats will like its tender branches, leaves and seeds.

142　长叶铁扫帚

143　截叶铁扫帚

学名　*Lespedeza cuneata*（Dum.-Cours.）G. Don
英名　Cuneate Bushclover

别名铁扫帚、绢毛胡枝子。多年生草本。茎直立，多分枝，枝细长，薄被微柔毛。三出复叶互生，密集，叶柄极短；小叶矩圆形、线状楔形，先端钝、截形或微凹，有小锐尖。总状花序腋生，有花1~4朵；小苞片2，卵形；花冠蝶形，黄白色或乳白色，有紫斑；子房上位，花柱内曲，柱头小，顶生。荚果细小，无柄，薄被丝毛。内含种子1粒，成熟后不开裂。生于山坡、空旷荒地、路边或河谷灌丛之中。营养期枝叶较柔嫩，果期后粗老，适口性下降。植物体内含一定量的单宁，家畜开始不习惯采食，一经习惯即喜食。

A semi-shrub, reproduces from seed. It exists in hill slopes, field-sides and roadsides. It is distributed in Gansu, Shaanxi, Shandong and Henan, etc. It is an excellent forage, sheep goats and cattle like to graze when fresh.

143-1　截叶铁扫帚

143-2　截叶铁扫帚叶

144　达乌里胡枝子

学名　*Lespedeza davurica*（Laxm）Schindl
英名　Dahurian Bushclover

别名兴安胡枝子。小灌木。枝有棱，具短柔毛，老枝黄褐色或赤褐色，嫩枝绿褐色。小叶3，披针状长圆形或狭长圆形，叶面无毛，背面被伏贴短毛。总状花序腋生，稍密集，花冠黄绿色，有时基部紫色。荚果倒卵状矩形，被白色柔毛，有明显的网纹。达乌里胡枝子与截叶铁扫帚的主要区别是，小叶披针状长圆形，先端圆钝，萼齿先端为长刺毛状，略短于花冠；小花密集似头状。我国北方普遍分布，生于山坡、沟底、田埂或沙滩等无耕作区，各种家畜喜食其嫩枝叶，粗老后适口性下降。

A semi-shrub reproduced from seeds. It is commonly found in arid hill slopes, wastelands, roadsides, sand beaches, field-sides and orchards. It is distributed in provinces of north, northeast and northwest China as well as Yunnan province. The feeding value is high in the early season when livestock are willing to grazing its tender tissues.

144-1　达乌里胡枝子

144-2　达乌里胡枝子花序

十八 豆 科

145 天蓝苜蓿

学名　*Medicago lupulina* L.
英名　Black Trefoil

越年生或一年生草本。茎自基部分枝，枝多而铺散地面或半直立，有疏毛。三出复叶，小叶倒宽卵形至菱形，上部叶缘有锯齿，两面均有毛。托叶斜卵形，有柔毛。花10多朵密集成头状花序，花序梗细长。花萼钟状，萼筒短，萼齿长。花冠黄色，稍长于萼。荚果弯曲成肾形，无刺，内含一粒种子。种子倒卵形或肾状倒卵形，主要靠种子繁殖。在北方分布普遍，生于较湿润的地方，包括渠边、路旁、农田、果园和撂荒地。营养丰富，适口性好，各种家畜家禽均喜食。

A biennial or annual legume reproduced from seeds.It is commonly found in fairly moist fieldsides, roadsides, wastelands and farmlands.It is distributed throughout the Loess Plateau as well as southern China, such as Sichuan and Yunnan, etc.The feeding value is high and most domestic animals like to graze in four seasons.

145-2　天蓝苜蓿花序　　　145-1　天蓝苜蓿

146 扁蓿豆

学名　*Pocockia ruthenia*（L.）Boiss.
英名　Russian Fenugreek

别名花苜蓿、野苜蓿。多年生草本。根系发达。茎常四棱形，疏生短毛，长20~60厘米，斜生或平卧，多分枝；羽状三出复叶，小叶圆形或倒卵形，总状花序，腋生，着花4~10朵，荚果扁平，矩圆形，种子椭圆形，淡黄色。典型草原和沙性植被的伴生种，多生于沙地、丘陵、河岸等处。我国北方从东到西均有分布。草质优良，适口性好，各种家畜均喜食。高寒地区补播或建立人工草地的首选种。

A perennial legume, reproduces from seeds.It can resist cold and drought conditions, distributed north, northwest and northeast China.It is one of the best forage in that area, various domestic animals like to graze it.

146-1　扁蓿豆　　　146-2　扁蓿豆花序

147　花苜蓿

学名　*Medicago ruthenica* L.
英名　Russian Fenugreek

别名野苜蓿、扁苜蓿。多年生草本。茎近四棱形，疏被白色短柔毛，长可达1米左右，斜生，或直立，或平卧，多分枝。三出复叶，小叶多型，倒卵形、倒卵状楔形、卵圆形或细条椭圆形，先端圆形、截形或微缺，具小尖头，基部楔形，边缘有锯齿，上面无毛，下面有白色柔毛。小叶柄长3～5毫米，被柔毛，托叶披针形，先端尖。总状花序腋生，花小，多少不一。花萼

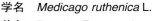

147-1　花苜蓿花序

钟状，花冠蝶形，黄色，有清晰的紫纹。荚果扁平，长圆形，每荚含种子2～4粒，种子卵形或矩圆形，黄褐色。我国主要分布在"三北"，生于丘陵坡地、河边沙地、山坡草地、农田、果园周边也有。为优质牧草，各种家畜均喜食。

147-2　花苜蓿

A perennial legume, it is often grows on hilly sloping fields, sands and on grasslands of roadsides in Loess Plateau. The feeding value is high. Most domestic animals like to graze it all year around. It has been found to stimulate milk production for dairy cows.

148　披针叶黄花

学名　*Cassia tora* L.
英名　Scckle sonna

别名假绿豆。一年生。形似半灌木状草本，羽状复叶，小叶6片，倒卵形或倒卵状矩圆形，在叶轴上两小叶之间有一腺体，幼嫩时两面疏生柔毛。花通常2朵生于叶腋，总花梗极短，花冠黄色。荚果条形，长10～20厘米，直径3～4毫米，种子多数。黄土高原南缘分布较多，一般生于海拔1 000米左右的山坡、路边，也入侵农田、果园、蔬菜地，江南有栽培种。牧草营养含量高，适口性好，家畜喜食。

A perennial legume, it is a salt-tolerant plant. It is distributed throughout the Loess Plateau. Cattle and sheep will graze in end of autumn and winter or drought year.

148-2　披针叶黄花花序　　**148-1　披针叶黄花**

149 洛氏锦鸡儿

学名　*Caragana roborovskyi* Kom.
英名　Desert Peashrub

别名荒漠锦鸡儿、猫耳刺、母猪刺。多年生灌木。高 0.3～1 米，直立或外倾，由基部多分枝。老枝黄褐色，被深灰色剥裂皮；嫩枝密被白色柔毛。羽状复叶有 3～6 对小叶；托叶膜质，被柔毛，先端具刺尖；叶轴宿存，全部硬化成针刺，密被柔毛；小叶宽倒卵形或长圆形，具刺尖，基部楔形，密被白色丝质柔毛。花梗单生；花萼管状，密被白色长柔毛；花冠黄色，旗瓣有时带紫色，倒卵圆形，基部渐狭成瓣柄，翼瓣片披针形；子房被密毛。荚果圆筒状，被白色长柔毛，先端具尖头，花萼常宿存。强旱生灌木，生长在荒漠、半荒漠草原、青藏高原和黄土高原的低山丘陵。家畜冬季重要的牧草，羊、骆驼、马采食，有保膘功效，特别在灾害严重的年份，它能为当地家畜发挥救命草的作用。

149-1　洛氏锦鸡儿

149-2　洛氏锦鸡儿叶序

A shrub, reproduced from seeds and by rhizome. It grows well on drought condition such as desert or hill slopes of the Loess plateau. It is distributed in Gansu, Ningxia, Xinjiang and Inner Mongolia. It is very important forage for camel in spring.

150 白皮锦鸡儿

学名　*Caragana Leucophloea* Pojark
英名　White-bark Peashrub

落叶灌木。高可达 2 米，树皮黄白色，枝条开展，小枝细长有棱，密被短柔毛。托叶和叶轴在长枝上硬化为针刺，宿存；小叶 4，无柄，顶端一对常较大，假掌状，狭倒披针形或条形，先端锐尖，顶端微凹，有短刺尖，淡绿色，有时带红色，疏被长柔毛或无毛。花单生于短枝叶丛中，花冠黄色或深黄色。荚果稍扁，条形，无毛。耐干旱，抗风沙能力强，多生长在极端严酷的生态环境之中。分布于我国甘肃和内蒙古西部、新疆等地。荒漠地区骆驼和山羊的饲用植物，骆驼一年四季喜食，山羊在春季喜食嫩枝。

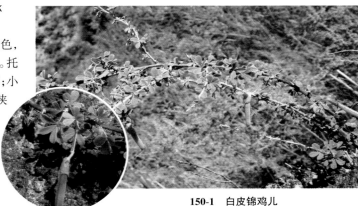

150-1　白皮锦鸡儿

150-2　白皮锦鸡儿荚果

A shrub, reproduced from seeds. It exists on very drought conditions, and distributed in northwest Loess plateau as well as Xinjiang and western Inner Mongolia. The feeding value is high and camel like to graze year around.

151　中间锦鸡儿

学名　*Caragana intermedia* Kuang et H. C. Fu
英名　IntermediatePeashrub

别名拧条。多年生丛生灌木。多分枝，树皮黄灰色、黄绿色或黄白色；枝条细长，幼时被绢状柔毛。托叶宿存，硬如针刺。羽状复叶，小叶先端钝圆或锐尖，有小刺尖。花梗长 8～12 毫米，常中部以上具关节；萼筒状钟形，密被短柔毛；花冠黄色，旗瓣宽卵形或菱形，翼瓣与龙骨瓣长圆形；子房披针形，无毛或疏被短柔毛。荚果披针形或长圆状披针形，顶端短渐尖。喜生于沙、砾质土壤，耐寒、耐酷热、耐瘠薄。产量高，适口性好，抓膘饲用植物。春季绵羊、山羊喜食嫩枝叶，骆驼终年喜食，马、牛不喜食。

A shrub reproduced from seeds.It mainly grows in northwestern Loess plateau.The feeding value is moderate.Sheep and goats will graze its tender branches and leaves in spring. Cattle use is limited.This plant can be used to make hay powder as a substitute for fodder.

151-1　中间锦鸡儿

151-2　中间锦鸡儿花序

152　小叶锦鸡儿

学名　*Caragana korshinskii*
英名　Korshinsk Peashrub

别名拧条，灌木。高可达 1 米，树皮金黄色，有光泽。枝条细长，小枝灰黄色或黄褐色，具条棱，密被绢状柔毛，长枝上的托叶宿存并硬化成针刺状，有毛。羽状复叶，小叶 12～16 片，倒披针形或矩圆状倒披针形，两面密生绢毛。花多为单生，花冠黄色。荚果圆筒形，略扁，革质，深红褐色，顶端渐尖。本种为耐旱喜沙植物，再生能力强，多生于草原地带的沙地、固定或半固定沙丘及山坡，分布在内蒙古高原和黄土高原生长条件较差的荒山秃岭。嫩梢山羊、绵羊喜食，花的营养价值高，有抓膘作用。骆驼终年喜食，牛、马采食较少。

152-1　小叶锦鸡儿花序　　152-2　小叶锦鸡儿

A shrub reproduced from seeds.It is cultivated in northwest China and the Loess Plateau.It is a good quality forage.Sheep,goats and camels like its tender branches and leaves in spring while horse and cattle graze limited.

153　二色棘豆

学名　*Oxytropis bicolor* Bge.
英名　Twocolor Crazyweed

多年生草本。茎极短,似无茎状。羽状复叶,小叶多对生,披针形,先端急尖,基部圆形,两面密生长柔毛。托叶卵状披针形,密被长柔毛,与叶柄连合。花多数,生于花序梗的上部,排列成较密的总状花序。花萼筒状,密生长柔毛,萼齿5,三角形。花红紫色或蓝紫色,旗瓣菱状卵形,腹面中央有绿褐色斑。荚果长圆形,密生长柔毛,2室。种子肾形或近方形,黄褐色,依靠种子繁殖。生于渠边、路旁、荒地、果园,家畜喜食,经霜后适口性提高。

153-1　二色棘豆

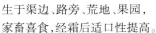

153-2　二色棘豆花序

A perennial legume that grows on hillsides, sandy grasslands and typical steppes. It is distributed in north, northwest and northeast China. The feeding value is moderate. Sheep and goats are willing to graze it when fresh or dry. Cattle and horses graze a little.

154　甘肃棘豆

学名　*Oxytropis kansuensis* Bunge
英名　Herb of Kansu Skullcap

多年生草本。高15~20厘米,基部有分枝,通体疏生白色长柔毛,间有黑色短柔毛。单数羽状复叶,叶轴上面具沟;托叶卵状披针形,基部连合,与叶柄分离;小叶13~25,卵状矩圆形至披针形,先端渐尖,基部圆形,两面有密长柔毛。总状花序近头状,腋生;总花梗长5.5~15厘米;花萼钟状,萼齿条形,与筒部近等长;花冠黄色。荚果长椭圆形或矩圆状卵形,膨胀。生于干燥草原、山坡、河边或林下。分布甘肃、青海、四川、云南等地。家畜较喜食。

A perennial legume, it is commonly found in hill slopes, grasslands, under forests or riverbed. It is distributed in Gansu, Qinghai, Sichuan and Yunnan provinces. The feeding value is low, but sheep and cattle will graze it sometimes.

154-1　甘肃棘豆　　154-2　甘肃棘豆花序

155 苦豆子

学名 *Sophora alopecuroides* L.
英名 Foxtail-like Sophora

别名狐尾槐。小灌木。茎直立,基部多分枝,密生灰白色平贴绢毛。奇数羽状复叶,小叶长圆状披针形或长圆形,先端尖或钝,基部近圆形或楔形,两面密生平贴绢毛,灰绿色。总状花序顶生,花密集。花萼钟状,有毛;花冠淡黄色。荚果串珠状,向内弯曲,有毛。种子椭圆形,黄色至黄褐色。常连片生长在阳光充足的山坡、路边、碱性土壤及沙丘。一般情况下,家畜不喜食。

155-1 苦豆子　　　155-2 苦豆子花序

A perennial legume, it is spread by crown buds and by seeds. It is commonly found in sunny calcareous soil or around sand dunes. It is distributed in north and northwest China. It is a toxic plant. Sometimes, animal will like its inflorescences.

156 苦马豆

学名 *Sphaerophysa salsula* (Pall.)DC.
英名 Salt Globepea

小灌木。茎直立或倾斜,有疏生倒伏毛。羽状复叶,小叶倒卵状椭圆形,先端微凹或圆形,基部近圆形或宽楔形,叶面无毛,背面有白色伏毛。托叶披针形,有白色伏毛。总状花序腋生,花萼杯状,萼齿5,有毛。花冠红色,旗瓣圆形,边缘向后卷曲,龙骨瓣比翼瓣长。荚果长圆形,膜质,膨胀成膀胱状,有长柄。种子肾状圆形,棕褐色。主要分布在华北、西北,生长于低湿沙地、肥沃荒地。其枝叶可以饲喂家畜。

A semi-shrub reproduced by seeds and crown buds. It is commonly found in river banks and moist low sandy lands. It is distributed throughout the Loess Plateau and in north China. Sheep, goats and camels will graze after it is withered. Anials do not graze when fresh.

156-1 苦马豆　　　156-2 苦马豆花序

157 三齿萼野豌豆

学名 *Vicia bungei* Ohwi
英名 Bunge Vetch

157-1 三齿萼野豌豆

157-2 三齿萼野豌豆花序

多年生草本。根系发达，茎四棱，柔弱，多伏地，少攀缘，分枝繁盛。羽状复叶，有卷须。小叶长圆形，先端截形或凹，具短尖，基部圆形，叶背疏生柔毛，托叶有齿。总状花序腋生，有花2~4朵。花萼斜钟状，萼齿5，宽三角形，上面2齿较短，疏生长柔毛。花冠蓝紫色或红蓝色。荚果长圆形，略膨胀，黄褐色。生于田埂、休闲地、路边、农田、果园或刚退耕的土地。属于优质牧草，各种家畜喜食。

A perennial legume, spread by crown buds and by seeds. It occurs in wastelands and grasslands, distributed in north China and the Loess Plateau. The texture of the plant is soft, with large amount of leaves, palatability is good and nutritive value is high. Livestock like it.

158 山野豌豆

学名 *Vicia amoena* Fisch. ex DC
英名 Wild Vefch

多年生草本，也有一年生或二年生者。茎四棱，攀缘或斜升。双数羽状复叶，具8~14片小叶，叶轴末端成分枝或单一的卷须。托叶披针形或戟形，小叶狭椭圆形或狭披针形，先端尖，基部圆，叶背面有短柔毛。总状花序腋生，与叶近等长，有花10~30朵，紫色或蓝紫色。花萼钟状，花冠蝶形。荚果长圆状菱形，两端尖，无毛，含种子2~4粒。种子圆形，黑色或褐色不一。我国南北广布，退耕地较多。营养生长与生殖生长几乎同时进行，边伸长，边现蕾，边开花，边结荚，边成熟，种子成熟快。在适宜的土壤、气候条件下，每667米^{2}鲜草产量可达5 000千克。营养物质含量高，草质柔软，适口性好，多种家畜喜食。

158-1 山野豌豆

158-2 山野豌豆花序

A biennial or perennial trailing herb reproduced from seeds. It occurs in river banks, lank banks, waste lands, grassy hill slopes and farmlands. It is distributed throughout the country. The feeding value is quite high with various animals like to graze.

159 小巢菜

学名 *Vicia hirsute* L.
英名 Pigeon Vetch, Tare Vetch, Hairy Vetch

别名硬毛果、野豌豆、雀野豆。一年生或二年生草本。无毛,细柔蔓生。偶数羽状复叶,小叶 8~16,有分叉卷须。小叶长椭圆形或矩圆形,截形或微凹,具细尖,托叶半戟形。总状花序腋生,有花 4~6 朵;花冠白色或淡紫色,花萼钟状,5 萼齿等长,线状针形。旗瓣椭圆形,顶端截形有细尖,翼瓣顶端圆形,和旗瓣等长,龙骨瓣比旗瓣稍短,子房被棕色硬毛,花柱顶端四周被柔毛。荚果矩圆形,两侧扁,被硬毛。每荚含种子 2 粒。广布于我国,喜生于温暖潮湿肥沃的石灰质土壤。茎叶柔嫩,适口性良好,各种家畜均喜采食,以牛、山羊和绵羊为最。

A biennial or annual creeping herb, reproduced from seeds. It occurs in grasslands around foot hills and roadsides, grows under shrubs and in crop The feeding value is quite high for various animals.

159 小巢菜

160 木 蓝

学名 *Indigofera tinctoria* L.
英名 True Indigo

别名小柴篮子。小灌木。茎高 50~60 厘米,被白色丁字毛。奇数羽状复叶,小叶 5~13 片,倒卵形或倒卵状短圆形,先端尖,基部宽楔形,全缘,两面均被丁字毛。总状花序腋生,多短于复叶。花蝶形,花冠红色。荚果圆柱形,黑褐色,有丁字毛。种子多数,矩圆形,褐色。主要生长在山坡、沟谷灌丛之中。放牧家畜喜食,食后容易上膘,无论青草还是干草都是优良的饲草。

A shrub, commonly found among shrubs, water side in gullies and roodsides. It is distributed in shanxi. Gansu, Helei, sichunn, jiangsu and Guangxi. It is one of the excellent forage. Cattle, horses, sheep and rabbits will graze it.

160-2 木蓝花序　　**160-1 木蓝**

161 合 欢

学名 *Albizzia julibrissin* Durazz.
英名 Silktree Albizzia

别名绒花树、马缨花、夜合树。落叶乔木。高 4～15 米。树皮灰褐色或淡灰色,不裂。小枝略带棱

161-1 合欢

161-2 合欢花序

角,有纵细纹,疏生皮孔。2 回羽状复叶,有羽片 4～12 对,小叶 10～30 对,长圆形至线形,两侧极偏斜。花序头状,多数,伞房状排列,腋生或顶生;花淡红色。荚果条状,扁平,褐色,边缘有厚棱。幼时有毛,每荚含种子 10 粒左右。适生于低山丘陵、沟谷坡地及林缘。叶量大,叶片柔软,营养丰富。幼嫩茎叶和荚果饲喂绵羊、山羊和牛效果好,叶粉是猪和家禽的优质饲料。

A woody plant reproduced from seeds. It is cultivated in south and central Loess Plateau as well as southern China. The feeding and nutritive value is quite high. Cattle, sheep and goats like to graze its leaves and seeds.

162 白花刺

学名 *Sophora viciifolia* Hance
英名 Vetchleaf Sophora

别名狼牙刺、苦刺、马蹄针。灌木。具锐刺,枝条和叶轴被柔毛。奇数羽状复叶,小叶 10～20 片,椭圆形或长倒卵形,先端钝或微凹,有短尖头,两面疏被短柔毛,中脉被毛较密。托叶针刺状,宿存。总状花序生于枝端,花 10 朵左右。花萼钟形,蓝紫色,被短柔毛。花冠白色或蓝白色,旗瓣倒卵形,反曲,龙骨瓣基部有钝耳。荚果串珠状,具长喙,近无毛,果皮革质,开裂,内含种子 1～5 粒,种子椭圆形,黄色或黄白色。主要分布在黄土高原及周边的太行山、秦岭北坡低海拔处,成片着生于草地和荒芜地。分枝多,幼嫩枝叶量大,牛、羊均喜食其凋落的干叶。

162-1 白花刺   162-2 白花刺花序

A shrub, reproduced from seeds and by rhizome. It is commonly found in hill slopes and wastelands, distributed in north and central Loess plateau. The feeding value is moderate or low, livestock animals like to graze its leaves only.

163　红花岩黄芪

学名　*Hedrysarum multijugum*
英名　Multijugate Sweetvetch

半灌木。高 50~100 厘米。茎有白色柔毛，下部木质化。奇数羽状复叶；小叶 11~35，卵形、倒卵形或宽椭圆形，上面无毛，下面有白色短柔毛；小叶柄短；托叶三角形，膜质。总状花序腋生，花疏；花萼斜钟状；花冠红色或紫红色，子房有柔毛。荚果扁平；荚节 2~3，近圆形，有肋纹和小刺，具白色柔毛。中旱生植物，适应性强，喜光、抗寒、抗风沙，生于沙质草原、固定的沙丘、沟坡、崖畔，主要分布在西北。

163-1　红花岩黄芪　　　　　　　　163-2　红花岩黄芪花序

属于优良饲用植物，无论干、鲜，马、牛、羊、兔均喜食。

It grows on hills of grasslands and in sands, and resists to cold and drought, and is distributed in northwest Loess Plateau as well as many parts of Mongolia, Xinjiang, Qinghai. It is good forage and the feeding value is high, grazed by all livestock when fresh or dry.

164　美丽胡枝子

学名　*Lespedeza formosa* (Vog.) Koehne
英名　Spiffy Bushclover

落叶灌木。高 1~2 米，干皮黑褐色，有细纵棱，幼枝密被白色短柔毛。三出复叶，小叶椭圆状披针形或卵状椭圆形，先端急尖、钝圆或微凹，背面密被白柔毛，总状花序腋生，单生或排成圆锥状，总花梗及小花梗均被白色柔毛，花紫红色，两性；花萼被毛，萼裂长于萼筒。荚果卵形或矩圆形，稍偏斜，先端有短尖，被锈色短柔毛。适应性广，耐旱，常生于丘陵山地的林缘和灌草丛中。分布于甘肃、陕西、河北、山西、山东、河南等省。枝条柔嫩，叶量大，属中等牧草，牛、羊均采食。

A small semi-shrub reproduced from seeds. It is commonly found in field ridges and roadsides, distributed over parts of the Loess plateau. The feeding value is high and all livestock lick to graze it when it is immature.

164　美丽胡枝子

165 蓝花棘豆

学名 *Oxytropis coerulea* (Pall.) DC.
英名 Herba oxytropis falcatae

多年生草本。根系发达，入土深，近无茎。奇数羽状复叶，小叶 17～41，对生，卵形、矩圆形或卵状披针形，先端急尖或钝，基部圆形，两面疏生平贴长柔毛。托叶条状披针形，膜质，基部与叶柄合生；叶轴细弱，疏生短柔毛；花多数，排列成疏生的总状花序；总花梗比叶长或近等长，疏生短柔毛；花萼钟状，有密短柔毛，萼齿披针形；花冠蓝色、紫蓝色、紫色、深紫色或白色。荚果卵状披针形，膨胀，疏生白色及黑色短柔毛。多生长在土层薄、岩石裸露的砾石坡地、干旱山坡，分布于华北、西北。适口性好宜放牧，被视为抓膘牧草。

A perennial legume, commonly found in sandy lands, typical steppes and hill-slopes. It is distributed in Gansu, Shaanxi, Ningxia, Shanxi, Hebei and Inner Mongolia. It is an excellent forage for fattening and cattle, sheep and goats will graze it.

165-1 蓝花棘豆

165-2 蓝花棘豆花序

166 青海黄芪

学名 *Radix Astragali*
英名 MilkvetchRoot

多年生草本。根系发达，主根粗壮，入土深达 1 米以上。茎匍匐，长 10～40 厘米，上部有分枝，通体被白毛。奇数羽状复叶，小叶 11～17 枚，广椭圆形、椭圆形或倒卵形，先端圆形或截形，有细尖。托叶小，离生。总状花序腋生，花轴长 1～2 厘米，有 1 或数朵蓝紫色花。萼钟状，外面杂有少数黑毛，里面无毛。具 5 萼齿；子房有长柄。荚果膜质，半卵圆形，无毛。耐寒、耐旱、耐践踏。常见于高海拔砾石山坡，广泛分布于青藏高原，是我国特有的牧草品种。茎细、叶多，适口性好。高蛋白、低纤维、营养丰富。

A perennial legume, reproduces from seeds. It can grow at high altitudes and resist heavy use, cold and drought conditions. It is mainly distributed in Qinghai province and around. The feeding value is high, when it is fresh, sheep, goats and cattle are willing to graze and it can be used as hay.

166-1 青海黄芪花序

166-2 青海黄芪

167 毛胡枝子

学名 *Lespedeza tomentosa*（Thunb）Sieb

别名山豆花。小灌木。通体被白色柔毛。茎自基部分枝，斜生或近直立，三出复叶，互生，具短柄；小叶长圆形或卵状长圆形，先端圆形，有短尖。托叶条形，有毛。总状花序腋生，花序梗较长，花密集；花萼浅杯状，萼齿5；花白色或浅黄色，旗瓣基部中央为紫红色。荚果倒卵状椭圆形，有白色短柔毛。生于沙质地、向阳山坡或灌草丛中，荒地、路边和田埂也多见，广布我国西北和中东部。优良牧草，大家畜和放牧家畜尤为喜食。

167-1 毛胡枝子

A semi-shrub reproduced from seeds.It occurs in sandy lands, sunny grass slopes, farmlands and orchards. It is distributed in northeast China as well as Gansu, Shaanxi, Shanxi, Hebei and other northern provinces.It is an excellent forage, various animals will graze it.

167-3 毛胡枝子茎

167-2 毛胡枝子花序

十九 酢浆草科

168 酢浆草

学名 *Oxalis corniculata* L.
英名 Creeping Woodsorrel

别名三叶酸。因茎叶有酸味，所以也叫酸味草。多年生草本。全体通常被疏柔毛，茎匍匐或斜上，节上生根。三出复叶互生，叶柄细长，小叶倒心形，无柄。伞形花序腋生，总花序梗与叶柄等长。萼片5，长圆形，有毛。花瓣5，黄色，倒卵形。蒴果近圆柱形，有5棱，被短柔毛。种子椭圆形或卵形，褐色，种子繁殖。几乎分布全国，生于较潮湿的荒地、农田、地埂、路旁。家畜皆食，猪和兔最喜食。

A perennial herbaceous plant, with sour taste for its stems and leaves.The whole plant covers with sparse hairs with creeping stem and stolons.reproduced from seeds.It is distributed all over the country, mostly found in wetland and wasteland, in crops and roadsides.All livestock like it, especially pigs and rabbits.

168-2 酢浆草花序　　**168-1 酢浆草**

169 铜锤草

学名 *Oxalis corymbosa*
英名 Corymb Wood Sorrel

别名红花酢浆草。多年生草本。通常只有花葶而无茎。具肉质根，根上部有极易脱离的小鳞茎。基生三出复叶，长叶柄，上面疏生柔毛。小叶倒卵形或宽椭圆形，先端凹缺，两面有棕红色瘤状小腺体，被疏毛。伞房花序基生，长于叶，有花

169 铜锤草

5～10朵。萼片5，花瓣5，花瓣淡紫红色。蒴果短条形，具毛。小鳞茎和种子均可繁殖。生长在水肥条件较好的山坡草地，有时入侵蔬菜、花生地。略带酸味，家畜较喜食。

A perennial erect stem-less herb, propagates by bulbs and by seeds. It is commonly found in farmlands and wastelands, distributed in southern Loess Plateau as well as some provinces in southern China, such as Guangdong, Guangxi and Fujian. The feeding value is low with limited grazing for cattle and sheep, but the pigs are readily to graze after boiling.

170 黄花酢浆草

170-1 黄花酢浆草

170-2 黄花酢浆草花序

学名 *Oxalis corniculata* L.
英名 Creeping Woodsorrel

别名酸味草。多年生草本，种子繁殖。茎枝幼嫩细弱，匍匐或斜升，茎具节，节落地生根。三出复叶互生，茂盛；叶柄细长，小叶倒心形，无柄。伞形花序腋生，总花序梗与叶柄等长。萼片5，长圆形，有毛。花瓣5，淡黄色，倒卵形。蒴果近圆柱形，有5棱，被短柔毛。种子椭圆形或卵形，褐色。生于较潮湿的荒地、农田、地埂、路旁。分布几遍全国，南多北少。多种家畜采食，猪较喜食。

A perennial herb reproduced from seeds. It is commonly found in gullies and fairly moist wastelands in south of the Loess Plateau as well as southern China It is a good forage. Sheep and goats like it fresh or dry.

二十　牻牛儿苗科

171　牻牛儿苗

学名　*Erodium stephanianum* Willd
英名　Common Heron's Bill

171-1　牻牛儿苗　　171-2　牻牛儿苗花序

171-3　牻牛儿苗蒴果

别名太阳花。一年生或越年生草本。茎自基部分枝，平铺地面或斜升。叶对生，具长柄。叶片2回羽状深裂，小裂片条形，全缘或有少量粗齿。伞形花序腋生，总梗细长，通常有2~5朵花，萼片长圆形，先端有长芒。花瓣5，淡紫蓝色，上有深紫色条纹。蒴果先端有长喙，成熟时5个果瓣与中轴分离，喙部呈螺旋状卷曲。种子条状长圆形，褐色。多生长在旱作农田或初荒芜地上，幼嫩时各种家畜都喜食，果实长喙容易黏结羊毛。

A biennial or annual herb reproduced from seeds. It is commonly found in farmlands, wastelands and roadsides. It is distributed in the provinces and autonomous regions of the northern part of the country and around the Yangtze River valley. Livestock grazing is limited only after it is dry in autumn.

172　鼠掌老鹳草

学名　*Geranium sibiricum* L.
英名　Siberian Granesbill

172-2　鼠掌老鹳草花序

多年生草本。茎自基部分枝，常平卧于地面，有倒生疏柔毛。叶对生，掌状3~5深裂，裂片羽状分裂或成齿状深缺刻，两面均有毛。花单生于叶腋，花梗细长，果期常向侧弯曲。苞片2，披针形，生于花梗的近中部。萼片边缘膜质，先端有芒尖。花瓣5，白色，有深紫色条纹。蒴果具微柔毛，种子长圆形。遍布我国"三北"地区，生于水肥较好的地方，包括农田、果园、荒地等，放牧家畜喜食。

A perennial herb reproduced from seeds. It is commonly found in fairly moist wastelands and roadsides. It is distributed in provinces of north, northeast and northwest China as well as Sichuan and Hubei. The palatability is better than that of Erodium stephanianum.

172-1　鼠掌老鹳草

二十一 蒺藜科

173 蒺藜

学名 *Tribulus terrestris* L.
英名 Puncturevine Caltrap

一年生草本。通体被绢毛。茎自基部分枝,细长,平卧地面。偶数羽状复叶,互生。小叶片长圆形,全缘,柄极短。托叶披针形,小而尖。花单生于叶腋,5 萼片宿存。花瓣 5,黄色。雄蕊 10,5 长 5 短。果实为分果,由 5 个果瓣组成,果实成熟后分离。每个果瓣有长短刺各一对,且有硬刺和瘤状突起,内含种子 2~3 粒。以路旁多见。幼嫩时家畜喜食,花果期粗老,适口性下降,特别是果实容易黏结羊毛,刺伤口腔或皮肤。

An annual herb reproduced from seeds. It is commonly found in farmlands and on barren hillocks, and distributed throughout the Loess Plateau. Sheep and goats like to graze when it is fresh in spring. If the seeds matured, it is very harmful for sheep.

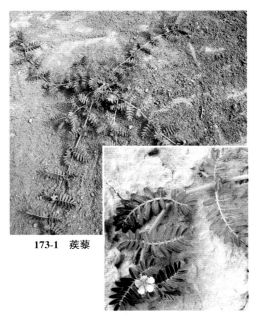

173-1 蒺藜

173-2 蒺藜花序

174 蝎虎霸王

学名 *Zygophyllum mucronatum* Maxim
英名 Crab Zygophyllum

别名念念、蝎虎草、蝎虎驼蹄瓣。多年生草本。茎多数,平卧或开展,有时具粗糙皮刺。小叶 2~3 对,肉质,条形或条状矩圆形,先端具短尖头。叶轴有翼,扁平,有时与小叶等宽。花腋生,1~2 朵,直立,花具梗,萼片 5,绿色,边缘白色,矩圆形;花瓣 5,白色或粉红色,稍长于萼片。蒴果披针形、圆柱形或窄卵形,弯垂,具 5 长棱,5 心皮,每室有 1~4 粒种子。种子椭圆形或卵形,黄褐色,表面有密孔。

该草为我国特有种,分布于甘肃、宁夏一带,生长在海拔 1 200~2 500 米,年降水 150~250 毫米的地区。适口性较差,马、牛不喜食,山羊、绵羊和骆驼乐于采食。

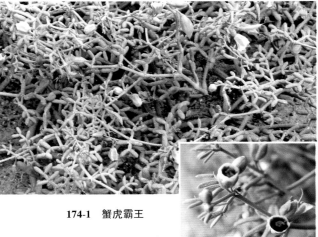

174-1 蟹虎霸王

174-2 蟹虎霸王花序

It is a strong xerophilous shrub, often grows on desert, steppe desert, rocky sloping lands. It is distributed over northwestern China. The feeding value is moderate. Camels, sheep and goats will graze its young branches and leaves.

175 白 刺

学名 *Nitraria tangutorum*
英名 Tangut Whitethorn

别名酸胖、泡泡刺、白茨。灌木。分枝多，平卧，先端针刺状。叶片倒披针形或宽倒披针形，先端钝圆或平截，全缘，通常2～3片簇生。聚伞花序生于枝顶，花多数。萼片5，绿色。花瓣5，白色。核果卵形或椭圆形，成熟时深红色，果核窄卵形，先端短渐尖。生长于沙漠地带。骆驼终年采食，羊于夏秋季采食嫩枝，果实是喂猪的优质饲料。

175-1 白刺

175-2 白刺花序

A woody shrub, spread by seeds and by crown buds. It is commonly found in sand dunes and sandy lands, distributed around north and northwest Loess Plateau. The feeding value is high and camels and sheep graze it all year around, but horses and cattle do not graze.

176 唐古特白刺

学名 *Nitraria Tangutorum* Bobr
英名 Tangut Nitraria

多年生落叶灌木。根系发达。多分枝，铺散地面，弯曲，有时直立。枝上可生不定根；小枝灰白色，先端针刺状。叶肉质，无柄，4～6片簇生，倒披针形，先端锐尖或钝，基部楔形，无毛或嫩时被柔毛。聚伞花序生于枝顶端，被疏柔毛；萼片5，绿色；花瓣白色，矩圆形。果实近球形或椭圆形，

176-2 唐古特白刺茎叶

两端钝圆，熟时暗红色。生于沙漠或沙漠边缘，骆驼喜食。

A shrub, It is commonly found in sandy soil, distributed in northwest Loess plateau as well as Xinjiang and Inner Mongolia. Sheep will graze its tender branches and leaves, and camel likes it year around.

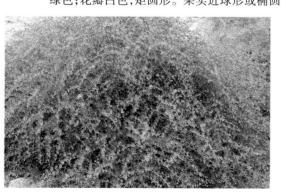

176-1 唐古特白刺

177　蒺根骆驼蓬

学名　*Peganum nigellastrum* Bunge
英名　Little Peganum

别名骆驼蒿。多年生草本。高 10~25 厘米，全株密被短硬毛。茎直立或开展，由基部多分枝，叶近肉质，2~3 回羽状全裂，小裂片条形，先端渐尖，托叶披针形。花较大，单生于分枝顶端或叶腋；萼片 5，披针形，各具 5~7 条状裂片；花瓣 5，白色或淡黄色；雄蕊 15，花丝基部加宽；子房 3 室。蒴果近球形，成熟时黄褐色，3 瓣裂。种子黑褐色，纺锤形，表面具小疣状突起。低等饲用植物，只有干草时，家畜采食少许。

177-1　蒺根骆驼蓬

177-2　蒺根骆驼蓬花序

A perennial herbaceous plant, spread by crown buds and by seeds. It is commonly found in desert steppes, hill slopes of arid area. It is distributed in northwest of the Loess plateau and Inner Mongolia. The feeding value is low. Camel, sheep and goats have limited grazing in winter.

178　骆驼蓬

学名　*Peganum multisectum* L.
英名　Multifid Peganum

多年生草本。根粗壮，肥厚而长，褐色。分枝多，分枝铺散，光滑无毛。叶互生，肉质，3~5 全裂，裂片条状披针形，有时顶端分裂。花单生，较大，花瓣 5，白色或淡黄色；雄蕊 15；子房 3 室，花柱 3。蒴果近球形，黑褐色，3 瓣裂，种子三棱形，黑褐色。低等牧草，青绿时骆驼采食，马、驴、羊少食或不食，霜降后羊和大家畜乐食。调制成干草对越冬抗灾有重要意义。

178-2　骆驼蓬花序

178-1　骆驼蓬

A perennial herb reproduced from seeds. It is commonly found mainly in drought grasslands and saline wastelands, found commonly in field-sides and roadsides. It is distributed in the provinces and autonomous regions of northwest China and the northern parts of the country. The feeding value is low but camels and goats like to graze it.

二十二 远志科

179 细叶远志

学名 *Polygala tenuifolia* Willd.
英名 Thinleaf Milkwort

多年生草本。主根粗壮发达。茎直立,基部分枝斜升,上部分枝开张,微被柔毛。叶互生,狭披针形,先端较钝,全缘。总状花序生于茎的上部,假顶生,通常有一两个花序高出全株。花排列稀疏,3苞片披针形,易落。萼片5,其中两片花瓣状;花瓣3,兰紫色。蒴果倒心形,扁平,边缘有狭齿。种子倒卵形,黑色,边缘密生白毛。多生于山坡草地、田埂,呈零散分布,山羊、绵羊均喜食。

A perennial herb, spread by seed and by crown bud. It is commonly found in sunny hill-slopes, field ridges and grasslands, distributed in the area north to the Yangtze River valley and east to Gansu province. It is excellent forage and it is palatable for livestock.

179-1 细叶远志

179-2 细叶远志花序

180 西伯利亚远志

学名 *Polygala sibirica* L.
英名 Siberian Milkwort

多年生草本。茎细直,光滑无毛,稍木质,分枝少。叶片披针形,先端尖,较细叶远志膜质。喜生于林缘、山地灌丛及向阳坡地、田埂。枝条细弱,柔嫩,山羊、绵羊均喜食。

A perennial herb, spread by seed and by crown bud. It occurs in sunny hillslopes, gras lands and sandy lands, distributed in north, northeast and northwest China. The feeding value is high and it is palatable for all grazing animals.

180-2 西伯利亚远志花序

180-1 西伯利亚远志

二十三 大戟科

181 铁苋菜

学名 *Acalypha australis* L.
英名 Copperleaf, Threeseed Mercury

别名杏仁菜、小耳朵草。一年生草本。茎直立，有分枝。叶互生，具长柄，叶片椭圆形、椭圆状披针形或卵状菱形，边缘有钝齿，两面无毛或有疏柔毛。花单性，雌雄同序，无花瓣。雄花序生于花序上部，穗状；雌花序着生于下部的叶状苞片内。蒴果小，钝三角形，灰褐色，有毛。种子倒卵球形，常有白膜质状的蜡层。分布于黄河流域。猪、兔、牛、羊、鹅等家畜均喜食，幼嫩和经霜后，适口性好转。

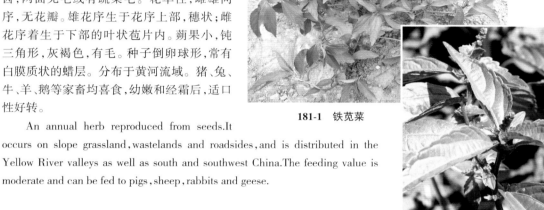

181-1 铁苋菜

181-2 铁苋菜花序

An annual herb reproduced from seeds. It occurs on slope grassland, wastelands and roadsides, and is distributed in the Yellow River valleys as well as south and southwest China. The feeding value is moderate and can be fed to pigs, sheep, rabbits and geese.

182 地 锦

学名 *Euphorbia humifusa* Willd.
英名 Humifuse Euphorbia

别名红丝草。一年生草本。全株有白色乳汁。植株枝叶纤细，叉状分枝，茎枝红色或淡紫红色，匍匐紧贴地面。叶对生，近无柄，叶片长圆形，先端钝圆，基部偏斜，边缘有细锯齿或近全缘，两面无毛或疏生柔毛。花序单生于叶腋，雌雄同序，无花被。总苞倒圆锥形，淡红色，先端4裂，蒴果三棱状球形。种子四棱状倒卵形，黑褐色，有蜡粉。该草遍布我国，北方较多，适应性广，到处可以生长，但多见于农田和荒地。山羊和绵羊均喜食，因枝细叶小而又紧贴地面使放牧家畜采食困难。

An annual prostrated herb reproduced from seeds. It is a common weed found in farmlands, wastelands and orchards. It is distributed throughout the provinces and autonomous regions in the Loess Plateau. It is moderate forage for sheep and goats.

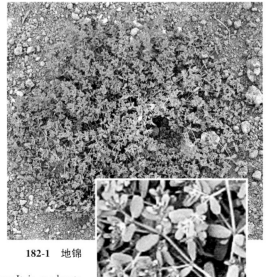

182-1 地锦

182-2 地锦花序

183　地构叶

学名　*Speranskia tuberculata*
英名　Speranskia

别名地构菜。多年生草本。根茎横走,淡黄褐色。茎直立,丛生,被灰白色卷曲柔毛。叶互生,无柄或具短柄;叶片披针形至椭圆状披针形,厚纸质,先端钝尖或渐尖,金缘,基部阔楔形或近圆形,下 2/3 部分具稀大齿,两面被白色柔毛,下面具腺体。总状花序顶生,密被短柔毛;花小,单性,同株;雄花具长卵状椭圆形或披针形的叶状苞 2 枚,苞片内具 1~3 朵花。萼片 5,稀 4;花瓣 5,稀 4,呈鳞片状。雌花具较长的花梗;萼片 5~6;花瓣 6。蒴果三棱状,顶端开裂。每室有种子 1 粒,三角状倒卵形,绿色。生于草地及干旱山坡,家畜不喜食。

A perennial herb, reproduced from seeds. It occurs in dry hill slopes, wastelands and field-sides, distributed over parts of the Loess plateau. The feeding value is low, livestock animals dislike to graze.

185-2　地构叶花序　　　185-1　地构叶

二十四　柽柳科

184　红柳

学名　*Tamarix ramosissima* Ledeb
英名　Branchy Tamarisk

别名多枝柽柳,灌木或小乔木。一般株高 2~3 米,多分枝,枝条紫红色或红棕色。叶片披针形、卵状披针形或三角状披针形,先端锐尖,略内弯。总状花序生于当年枝条上,组成顶生的大型圆锥花序。苞片卵状披针形,花梗短,萼片 5,卵形。花瓣 5,倒卵形,淡红色或紫红色。花盘 5 裂,雄蕊 5,花柱 3,棍棒状。蒴果长圆锥形,3 瓣裂。种子顶端簇生柔毛。广泛分布于西北沙漠边缘和大型盆地边缘,是我国干旱地区饲养骆驼的重要饲料。秋后山羊和绵羊采食其脱落的细枝,其余时间,其他家畜多不采食。

A shrub, it grows on lowlands between sand dunes on both sides of river in north and north-west China. It is good feed for camels to restore fat in early spring.

183-1　红柳

183-2　红柳花序

185　枇杷柴

学名　*Reaumuria soongorica*
英名　Songory Reaumuria

别名红砂、红虱。小灌木。多分枝。叶肉质，圆柱形，常3~5簇生，先端钝，浅灰绿色。花单生于叶腋或在小枝上集生为稀疏的穗状花序。苞片3，披针形，花萼钟形。花瓣5，粉红色或淡白色。蒴果长圆形，3瓣开裂，含种子3~4粒，种子长圆形。主要分布在我国的陕、甘、宁、蒙、新等省区，耐旱、耐盐，能适应强烈的干旱条件。羊、驼采食，马少食，牛不食。在饲草缺乏的年份，枇杷柴的饲用价值提高，是家畜主要的度荒草。因其盐分含量高，所以采食枇杷柴可以代替家畜补盐。

184-1　枇杷柴　　　　184-2　枇杷柴花序

A woody plant in desert area, distributed in Gansu, Xinjiang and Qinghai provinces. Camels like to graze fresh or dry all year around, it is a important forage for sheep in spring and winter, cattle usually do not graze.

二十五　胡颓子科

186　沙　棘

学名　*Hippophae rhamnoides* L.
英名　Chinese Seabuckthorn

多年生灌木。顶生或侧生许多粗壮直伸的刺。幼枝密被银白色带褐锈的鳞片，呈绿褐色；老枝灰褐色。单叶互生，有短柄。叶片条形或条状披针形，先端钝，基部圆。叶面绿色，幼嫩时被白色柔毛；叶背密被银白色鳞片。无花梗，花萼筒囊状。浆果球形或卵形，黄色，橘黄色或橘红色。种皮与果皮离生，种子倒卵形、倒阔卵形或倒楔形，淡褐色或棕褐色，顶端平截。山羊、绵羊均喜食枝叶。

186-1 沙棘

A shrub, commonly found round forest edge, grasslands and hill-slopes. It is distributed in northwest China. The feeding value is low and goats like to graze tender branches and leaves, other animals graze like its leaves in the winter.

186-2　沙棘浆果

187 沙 枣

学名　*Elaeagnus angustifolia* L.
英名　Russin olive

187-2　沙枣花蕾

187-1　沙枣

灌木或乔木。树皮栗褐色至红褐色,具枝刺,嫩枝、叶、花、果均被银白色鳞片及星状毛;叶具柄,披针形,先端尖或钝,基部楔形,全缘,上面银灰绿色,下面银白色。花小,银白色,芳香;通常1~3朵生于小枝叶腋;花萼筒状钟形,顶端通常4裂。果实长圆状椭圆形。多生于荒漠或半荒漠地区。各种家畜均喜食其嫩枝、叶及果实。

A big shrub, commonly found on migratory crescent dunes and sandy hill lowlands. It is distributed in north and northwest of the China. The leaves and fruits are very important forages. Cattle, sheep and camel will graze it in winter.

二十六　五加科

188 五 加

学名　*Acanthopanax giacilistylus* Smith
英名　Slenderstyle Acanthopanax

落叶小灌木。茎直立,有纵棱,带紫红色,分有刺和无刺两种。叶互生,掌状五出复叶,小叶卵形,先端尖,边缘上部有锯齿,叶面深绿,叶背淡绿。伞形花序,花黄绿色,浆果黑色。生于海拔500~1 500米的山坡灌丛之中,西北秦岭及余脉山林多见。幼嫩和秋后多种家畜均喜食,是一种优质饲料,夏季因具其特有的气味,家畜不喜食。另外,该植物也是一种传统的优质野菜。

A perennial shrub, commonly found among shrubs and grasslands, distributed in south and southeast Loess Plateau as well as other provinces such as Sichuan and Qinghai. The tender branches and leaves are good forages in autumn, and it is also edible in early spring.

188-2　五加花序

188-1　五加

二十七　葫芦科

189　栝　楼

学名　*Trichosanthis kirilowii* Maxim.
英名　Snakegourd Fruit

别名天撤、苦瓜、山金匏、药瓜皮,亦称瓜蒌。多年生攀缘草本。块根圆柱状,肉质,粗大,深入地下,富含营养且具再生功能。茎细而长,有纵棱。植株具2~5叉状的卷须。叶互生,具柄,叶片宽卵状心形,常3~7浅裂至深裂,边缘常再分裂,小裂片较圆,两面稍被毛。雌雄异株,雄花数朵形成总状花序,有时单生,萼片线形,花冠白色,裂片扇状倒三角形。雌花单生,花梗长。果实椭圆形至球形,随成熟逐渐变为橙黄色。种子扁平。生于山坡、谷地的草丛、林缘半阴处。中等牧草,果实、块根、茎叶家畜采食,亦可青贮。

A perennial trailing herb, spread from seed and by crown bud. It grows in hills, among thickets and under forest, distributed in north of the Loess plateau. Cattle, sheep, goats and horses like to graze its leaves, fruits, roots and tender branches.

189-1　栝楼

189-3　栝楼根

189-2　栝楼果

二十八　锦葵科

190　苘　麻

学名　*Abutilon theophrasti* Medic
英名　Chingma Abutilon Piemarker

别名白麻。一年生草本。茎直立,有柔毛。叶互生,具长柄。叶片心形,先端尖,边缘有粗锯齿,两面均有毛。花单生于叶腋,花梗细长。花萼杯状,5裂,花瓣5,鲜黄色。蒴果半球形,分果片多数,上面有粗毛,先端具2长芒。种子肾状卵形,灰褐色,有毛。在果园、薯类、豆类等农田中可见,低等牧草,牛、羊采食。

An erect annual herb reproduced from seeds. It is commonly found in farmlands, roadsides and wastelands, distributed in south, central Loess Plateau and throughout the southern China. The feeding value is moderate the livestock like to graze when young or withered.

190-1　苘麻　　**190-2　苘麻花序**

191 野西瓜苗

学名 *Hibiscus trionum* L.
英名 Flowerofanhour

别名香铃草。一年生草本。主茎直立,基部分枝常铺散地面,整个植株被白色星状粗毛。叶互生,具长柄。叶片掌状深裂或全裂,裂片倒卵形,通常羽状分裂,两面有星状粗刺毛。花单生于叶腋,小苞片12枚,条形。花萼钟状,裂片5,膜质,有绿色条棱,棱上有紫色疣状突起。花瓣5,白色或淡黄色,内侧基部呈紫色。蒴果长圆状球形,种子肾形,黄褐色,上面有瘤状突起物。我国各地都有分布,主要生长在荒地或豆类、瓜类植物中。属于中下等牧草,牛、羊等家畜较喜食。

191-1 野西瓜苗

191-2 野西瓜苗花序

An annual herb reproduced from seeds. It occurs in farmlands, roadsides and wastelands, and distributed throughout the country. The feeding value is moderate. The palatability is good. Horses and sheep like to graze it, and cattle grazing is limited.

192 圆叶锦葵

学名 *Malva rotundifolia* L.
英名 Running Mallow

多年生草本。全株通体被星状毛。茎自基部分枝,平卧地面,有时茎先端斜向上。叶互生,具长柄。叶片圆肾形或半圆形,叶脉明显且向背部隆起,叶缘有细锯齿和若干波状浅裂。花单生或数朵簇生于叶腋,花梗细长。小苞片3,披针形。萼片5,卵形。花瓣5,淡蓝紫色,脉纹色深,先端有缺口。分果扁球形,分果片肾形。叶多枝嫩,家畜初食时喜食,但不愿较长时间采食。

192-1 圆叶锦葵叶 192-2 圆叶锦葵

A perennial herb reproduced from seeds. It grows among hassocks along roadsides and nearby village. It is distributed throughout the Loess Plateau. The feeding value is moderate with sheep and cattle like to graze when young and after frost.

二十八　锦葵科

193　蜀葵

学名　*Althaea rosea* (L.) Caran.
英名　Hollyhock

别名棋盘花、鸡冠花、水芙。二年生草本。茎直立,分枝,被刺毛。单叶互生,叶片纸质,近圆心形,5~7掌状浅裂,裂片圆形至三角形,中裂较大。托叶卵形,先端具三尖。花腋生、单生或近簇生,通常呈总状花序。具叶状苞片,被星状长硬毛。花萼钟形,5裂,裂片卵状三角形。花冠多色,有单瓣和重瓣之分,花瓣倒卵状三角形,先端微凹,基部渐狭。花柱分枝多数,被细微毛,果盘状。家畜采食幼嫩枝叶或干叶,适口性欠佳,冬春缺草时采食较好。

A erect biennial herbaceous plant reproduced from seeds.It occurs on nearby villages and roadsides of arid region.It is commonly distributed in north, northeast and northwest China.Cattle, sheep and goats will graze its leaves in autumn.

193-1　蜀葵花序

193-2　蜀葵

194　野葵

学名　*Malva verticillata* L.
英名　Cluster Mallow

别名冬葵。越年生草本。茎直立,有星状长柔毛。叶互生,肾形或圆形,掌状5~7裂,基部心形,裂片卵状三角形,边缘有不规则的锯齿,主脉5~7条,两面被疏糙毛或近无毛。叶柄长,被长白色柔毛,托叶有星状柔毛。花小,淡红色或淡粉紫色,常簇生于叶腋。小苞片3,有细毛。萼片杯状,5齿裂,花瓣5,倒卵形或三角状倒卵形,顶端凹入。蒴果扁圆形或扁球形。种子繁殖力强。适口性较好,家畜采食。

An biennial herb, reproduced from seeds.It occurs nearby village, wastelands and field-sides, distributed over parts of the Loess Plateau.The palatability is good, cattle sheep and goats like to graze fresh branches and leaves.

194-1　野葵　　　　　　　　　194-2　野葵花序

二十九 藤黄科

195 长柱金丝桃

学名 *Hypericum ascyron* L.
英名 Longstyle St. John's wort

多年生草本。茎直立,具4棱。单叶对生,矩圆状卵形或矩圆状披针形,先端锐尖或钝,基部圆形或心形,抱茎,有透明腺点,全缘。花顶生,聚伞状,有时单生;萼片卵形或倒卵形,花瓣黄色或金黄色,倒卵形或披针形,雄蕊5束,短于花瓣;子房5室;花柱5;基部或中下部合生。蒴果卵状圆锥形,种子多数,表面具小蜂窝状纹,一侧具细长翼。饲用价值中等。

An arid shrub, spread on sunny mountain side, grasslands, field-sides or desert area. It is distributed throughout the Loess Plateau. The feeding value is moderate. Sheep and goats like to graze tender branches and inflorescence. Horses and cattle do not graze.

195-1 长柱金丝桃

195-2 长柱金丝桃花序

三十 堇菜科

196 早开堇菜

学名 *Viola prionantha* Bge.
英名 Serrate Violet

多年生无茎草本。叶基生,叶片长圆状卵形或长圆状披针形,叶基部平截或稍圆,叶柄从地面伸出,上部两侧具狭翅或无,直立或斜升。叶中脉明显,叶缘具细齿,叶面密生白毛。淡紫色花,两侧对称,花葶管状,稍显上细下粗。春季萌发极早,孕蕾时花葶弯缩在叶片基部。开花时花葶直立,长度略超过叶子。主要着生在路旁、半阴荒地、田边及果园。放牧家畜喜食。

A perennial stem-less herb. It is commonly found in wasteland in hill-slopes or plains, distributed in north of China and south of Loess Plateau. It starts to grow very early, the sheep and goats will to graze this plant in spring and autumn.

196-1 早开堇菜

196-2 早开堇菜花序

196-3 早开堇菜根

三十 堇菜科

197　紫花地丁

学名　*Viola philippica* Car.
英名　Purpleflower violet

多年生草本。无茎,主根粗壮明显,基生叶有柔毛。叶片短圆状披针形,或三角状卵形,边缘有波状钝锯齿。花两性,淡紫色,每株有花葶3~8个,较早开堇菜略短。开春返青早,开花早。蒴果。种子球形。该草我国各地均有分布,着生于山坡草地、沟谷、路边及村庄院落周围,属中等牧草,宜放牧,家畜在早春或晚秋霜后喜食。

197-1　紫花地丁

197-2　紫花地丁花序

197-3　紫花地丁根

A perennial stem-less herb, commonly found in field-ridges, roadsides and nursery gardens, distributed in Shaanxi, Gansu provinces and north, south, east and central China. The feeding value is low and sheep and goats will graze it in early spring and after frost in autumn.

198　犁头草

学名　*Viola japonica* Langsd.

多年生草本。无茎,叶基生。与紫花地丁的不同处在于,叶片长三角状卵形,基部宽心形,叶柄上有狭翅。花稍大,花葶也长些。花梗中上部有2个条形苞片,萼片5,披针形,附属物上有钝齿。花瓣5,淡紫色,有深色条纹,蒴果长圆形。与紫花地丁生境基本相同,比紫花地丁更喜潮湿的土地条件,该草适宜放牧,家畜喜食。

A perennial herb, commonly found in damp field-sides, orchards and farmlands, distributed over Shaanxi, southern Gansu in Loess Plateau as well as other provinces in Hebei, Hunan, Jiangsu and Jingxi, etc. The feeding value is moderate and sheep and goats like to graze in early spring or later autumn.

198　犁头草

三十一 柳叶菜科

199 柳叶菜

学名 *Epilobium hirsutum* L.
英名 Hairy Willowweed

多年生草本。根茎粗壮而坚硬,簇生须根。茎直立,上部分枝,茎密生白色长柔毛或短腺毛。茎上部叶互生,无柄,略抱茎;下部叶对生。叶片卵状长椭圆形,两面都有长柔毛,边缘具细锯齿。花单生于叶腋,萼筒圆柱形,4裂片,披针形,外被毛。花瓣4,先端2浅裂,淡紫红色。蒴果圆柱形,种子长椭圆形,顶端具毛。生于溪水、河流两旁或潮湿土壤,适口性中等,秋季经霜后家畜喜食。

It is a perennial herb, spread by rhizome and by seed. It common grows in marshlands of river banks and lake banks, distributed in provinces of northeast China and Hebei, Shaanxi, Shanxi, Xinjiang, Sichuan, etc. Palatability is poor, cattle and sheep grazing is limited at end of autumn.

199-2 柳叶菜花序　　199-1 柳叶菜

200 月见草

学名 *Oenothera odorata*
英名 Fragrant Evening Primrose

别名待宵草。基生叶丛生,有短柄,茎生叶互生,有短柄或无柄。叶片条状披针形,有不规则的疏锯齿,两面有白色短柔毛。夏季开花,花黄色,单生于叶腋。蒴果圆柱形,略具四棱,长2~3厘米,种子棕色。开花前各种家畜均喜食,以山羊为最。种子可榨油,油饼是家畜的精饲料。

200-1 月见草　　200-2 月见草花序

It is a perennial herb that reproducing from seeds. It is distributed in a parts of south of Loess Plateau as well as southern China. The yields and feeding value is high. Horses, cattle, pigs, gooses and dunks will graze it, particular sheep and goats.

三十二 伞形科

201 野胡萝卜

201-2 野胡萝卜花序

学名 *Daucus carota* L.
英名 Wild Carrot

201-1 野胡萝卜

越年生或一年生草本。全体被粗硬毛，根肉质粗大，圆锥形，近白色。茎直立，单生。基生叶矩圆形，有柄，2~3回羽状全裂，小裂片条形至披针形，具鞘。茎生叶近无柄，小裂片细长。复伞形花序，顶生，紧凑，有长柔毛和刚毛相互黏连。总苞片多数，叶片羽状分裂或不分裂，花白色、淡黄色或淡红色。果实为双悬果矩圆形或长圆锥形，有刚毛和翅，翅上具短钩刺。基生叶丛和嫩枝是家畜、家禽的优质饲料。

A biennial herb reproduced from seeds. It is commonly found in wasteland, orchards and among thickets, distributed in provinces of southwest and southeast China as well as Shaanxi province, etc. The feeding value is high. Cattle, sheep, goats and pigs like it.

202 硬阿魏

202-1 硬阿魏

学名 *Ferula bungeana* kitag.
英名 hard-ferula

别名沙茴香。多年生草本。直根圆柱形，淡棕黄色。茎直立，多分枝，开展，具细棱，节间实心。基生叶具长柄，基部加宽成叶鞘，茎生叶部分或全部成鞘状；1回和2回羽片具柄，3回羽片羽状深裂，裂片通常互生，末回羽片先端具2~3裂齿，齿端具细尖。复伞形花序多数，通常成层，轮状排列，伞辐4~10，具细棱；总苞片1~4，条状锥形，边缘膜质；萼齿卵形；花瓣黄色，卵状披针形；小舌片短于花瓣的1/2；果实矩圆形，压扁，背棱3条，隆起，侧棱宽翅状。具浓烈的特殊气味，家畜多不采食，只有嫩枝叶干燥后微嗜。

A perennial herbaceous plant, spread by crown buds and by seeds. It occurs around sandy lands, sand dunes and in farmlands, and is distributed in north and northeast China as well as Gansu and Shaanxi provinces. Sheep can use when young and will graze after autumn frost.

202-2 硬阿魏花序

203 水 芹

学名　*Oenanthe javanica*（Blume）DC.
英名　Javan Weterdropwort

别名水芹菜。多年生草本。茎无毛,基部匍匐。叶互生,1~2回羽状复叶。基生叶三角状或三角状卵形,小裂片卵形至菱状披针形,长2~5厘米,宽1~2厘米,边缘有不整齐的尖齿或圆锯齿。基部叶柄长10厘米以上,上部逐渐变短,基部延伸成鞘而抱茎,复伞形花序顶生。总花梗长;无总苞片;伞辐6~10,小总苞片5~10,条形。花白色。双悬果椭圆形,果棱隆起显著。一种良好的青绿饲草。叶量大,茎叶柔嫩多汁,具芹菜香味,猪尤为喜食,切碎后可喂家禽。

A perennial herb, spread by stolon and by seed. It occurs in damp low lands, shallow water and rice fields, distributed throughout the country. The feeding value is moderate. Good feed for horses, cattle and sheep.

203-1　水芹

203-2　水芹花序

204 柴 胡

学名　*Bupleurum chinense* DC.
英名　Chinese Thorowax

多年生草本。基部分枝或不分枝,茎丛生或单生。直立,上部分枝开张且较多,呈"之"字形弯曲。基生叶易早枯,中部叶披针形或宽条状披针形,小而少。叶两面绿色,全缘,有平行脉。复伞形花序腋生或顶生。花鲜黄色,双悬果宽椭圆形,

204-1　柴胡

棱狭翅状,具翅毛。耐旱,生于山坡,地埂,新鲜时因有特殊气味,家畜不喜食,干燥后适口性有所增强。

A perennial herb reproduced from seeds. It is commonly found in wastelands, hill-slopes and grasslands and distributed in northern China. The feeding value is low. Some animals do not like to graze when it is fresh. Cattle and sheep can consume a little as hay or powder in winter.

204-2　柴胡苗

205 窃 衣

学名 *Torilis japonica*（Houtt.）DC.
英名 Common Hedgeparsley

一年生或多年生草本。全株有贴生的短硬毛，茎单生，具分枝。叶片2回羽状分裂，小叶狭披针形，有短柄。复伞形花序，花白色。双悬果矩圆形，多刺毛。生于山坡、荒地、路旁。适口性中等，幼嫩时家畜喜食，开花粗老后饲用价值下降。

An annual or perennial herb, reproduced from seeds. It grows in hilly slopes, roadsides and wastelands. Various domestic animals like to graze when it is fresh.

205-1 窃衣　　　　205-2 窃衣花序

206 藁 本

学名 *Ligusticum sinense* Oliv.
英名 Rhizoma ligustici

多年生草本。主根通常不明显，根茎呈不规则圆柱状或团块状，表面棕褐色，着生多数绳状支根，具芳香气。茎直立，通常单一，节间中空，具纵细纹，常下部带暗紫色。茎下部叶和中部叶有长柄，上部叶柄抱茎；2～3回三出羽状全裂，一回羽片4～6对，具长柄；二回羽片2～4对，具短柄或无柄；最终裂片卵形或菱状卵形，先端钝而具小尖头，边缘有少数缺刻状牙齿，齿顶端有小尖头，两面沿脉有短糙硬毛。复伞形花序顶生与腋生，伞幅6～19，有短糙硬毛；无总苞片或有1片而早落；小伞形花序具多花；花梗长；小总苞片8～10，钻形；萼齿不明显，花瓣5，白色，椭圆形。双悬果椭圆形，分生果稍背腹压扁，背棱突起，侧棱狭翅状。主要生长在田埂和山坡草丛之中。饲用价值中等，早春和晚秋羊最为喜食。

A perennial herb reproduced from seeds. It is commonly found in field-sides, hilly slopes and among thickets, distributed in Jiangxi, Hunan, Hubei and Sichuan, etc. The feeding value is moderate, Sheep and goats like to graze in early spring and late autumn.

206-2 藁本花序　　　206-1 藁本

207　石防风

学名　*Peucedanum terebinthaceum*(Fisch.) ex Turcz.
英名　*Peucedanum terebinthaceum*(Fisch)

多年生草本，种子繁殖。根圆柱形或近纺锤形，下部分歧。茎直立，基部披棕黑色纤维状叶柄残基。基生叶有长柄，茎生叶叶柄较短，叶片2～3回羽状全裂，终裂片披针形，全缘或具缺刻状牙齿，表面脉上有糙毛，背面无毛；茎上部叶叶柄成鞘状，抱茎，边缘膜质。复伞形花序不等长，花瓣白色，中脉带黄色。双悬果广椭圆形，背腹压扁，种子腹面平坦。生于干燥山坡、山坡草地、林缘、林下、林间路旁。分布于东北、华北及西北。优良牧草，幼嫩时绵羊、山羊、猪、兔均喜食。

A perennial herb reproduced from seeds. It is commonly found in field-sides, wastelands and nearby village, distributed in south and central Loess Plateau. The feeding value is high. Sheep, goats, rabbits and poultry like to graze in early spring and late autumn.

207-1　石防风苗

207-3　石防风花序

207-2　石防风

三十三　报春花科

208　狼尾珍珠菜

学名　*Lysimachia barystachys* Bunge.
英名　Loosestrife

别名狼尾巴花。多年生草本。根细，多分枝，根状茎横走，棕红色，节上有鳞片。茎直立，偶有短分枝，茎上部被柔毛。叶互生，无柄，叶片长圆状披针形、披针形至倒披针形，先端尖，基部狭，全缘，边缘稍外卷，两面疏被柔毛。总状花序顶生，花密集，常向一侧弯曲呈狼尾状。花萼近钟形，深裂，裂片长圆状卵形，外被柔毛，边缘膜质。花冠白色，深裂，裂片长圆形或卵状长圆形。蒴果近球形，种子多数，红棕色。多生长在草坡、沟边、路旁，也入侵农田。家畜采食，幼嫩枝叶蒸煮后可喂猪。

A perennial herb, commonly found in wastelands, roadsides and gullies, distributed over all parts south to the Yangtze River and Shaanxi province, etc. The feeding value is moderate. Horses, cattle and sheep will graze it and pigs like its leaves.

208-2　狼尾珍珠菜　　208-1　狼尾珍珠菜花序

209　羽叶点地梅

学名　*Pomatosace filicula* Maxim.
英名　Common Pomatosace

一年生或二年生草本。国家Ⅱ级重点保护野生植物。花葶高3~9厘米。叶基生,沿中脉疏被长柔毛,羽状深裂,裂片线形,全缘或具不整齐的疏齿;叶柄疏被长柔毛。伞形花序着生于花葶端;苞片线形,疏被柔毛,花萼杯状,5裂,裂片三角形,内面被微柔毛;花冠稍短于花萼,白色,坛状,喉部收缩且具环状附属物,冠檐5裂,裂片长圆形;雄蕊5,着生于花冠管的中上部,与花冠裂片对生;花丝极短,花药卵形,先端纯;子房下位,扁球形,有胚珠数枚;花柱短于子房,宿存;柱头头状。蒴果近球形。种子6~12粒。分布于甘肃、四川、青海及西藏的高山草甸草原、河滩沙地和山谷阴坡。

209-1　羽叶点地梅　　209-2　羽叶点地梅花序

An annual or biennial herbaceous plant reproduced from seeds. It is commonly found in mountain grasslands and meadows, and distributed in southwest Loess plateau, Gannan rangeland in Gansu province and Qinghai-Xizang plateau.

三十四　白花丹科

210　金色补血草

学名　*Limonium aureum*
英名　Golden Sealavander

别名黄花矶松、金匙叶草。多年生草本。根圆柱状,木质,粗壮发达。叶基生,矩圆状匙形至倒披针形,顶端圆钝,具短尖头,基部渐狭成扁平的叶柄。花序轴2至数条,自基部开始呈二叉状分枝,"之"字状弯曲。聚伞花序排列于花序分枝的顶端而形成伞房状圆锥花序,花序轴密生小疣点。苞片宽卵形,边缘膜质。花萼漏斗状,5裂,金黄色,三角形,先端具1小芒尖。花瓣橘黄色,干膜质,基部合生。蒴果倒卵状矩圆形,有5棱,包被于宿存的花萼内。幼嫩和干燥后家畜喜食,开花前后粗老,适口性差。

210-1　金色补血草
210-2　金色补血草花序

A perennial herb, it is a salt-tolerant plant. It grows in hill-slopes, and can resist cold, drought and alkali, distributed in north and northwest Loess Plateau. The feeding value is low with limited grazing for animals.

211　二色补血草

学名　*Limonium bicolor*（Bunge.）Kuntze
英名　Twocolor Sealavender

多年生草本。全株无毛，根粗壮。基生叶多数，茎一至数条从叶丛中抽出，多直立，少倾斜。茎上多花枝，开张，少或无叶。基生叶片大，匙形或倒卵状匙形，先端钝而具短尖，基部下延成狭叶柄。圆锥花序由多数聚伞花序组成，有不育小枝，花无梗，数朵着生于分枝顶端的一侧，苞片绿色，短于花萼。萼筒圆锥状具柔毛。裂片5，白色；花瓣5，黄色，基部合生。生于林间、草地、山坡、农田周边。家畜采食基部叶和嫩枝，干枯后牛、羊皆食。

A perennial herb, spread by crown buds and by seeds. It usually exists on low slopes and gullies, distributed in south, southeast and central Loess Plateau. The feeding value is moderate. Cattle, sheep and goats like to graze fresh leaves and inflorescences.

211-1　二色补血草

211-2　二色补血草花序

三十五　龙胆科

212　龙　胆

学名　*Gentiana squarrosa* L.
英名　Roughleaf Gentian

别名小龙胆、石龙胆、鳞叶龙胆。一年生矮小草本。高3～10厘米，茎短小，多分枝。叶对生，无柄，基部叶卵圆形或倒卵状椭圆形，上部叶匙形至倒卵形，先端具芒尖，反卷。花单生于茎顶，花冠钟形，5裂，裂片卵圆形，褶全缘或2裂，雄蕊5，着生于花冠筒中部。子房上位，花柱短。蒴果倒卵形，果梗长。我国"三北"地区均有，六盘山海拔2 000米左右阳坡较多，是莎草和蒿属植物的伴生种。适口性差，牛、羊微嗜。

A perennial small herb, it grows on steppes, meadow steppes and mountain steppes in north, northwest and northwest China. The feeding value is moderate. Cattle and sheep like to graze it.

212-1　龙胆　　212-2　龙胆花序

213 獐牙菜

学名　*Swertia bimaculata*（Sieb.et Zucc）
英名　Twospot Swertia

多年生草本。茎直立,暗红色,有钝纵棱,中上部分枝,单生或对生,新生茎色淡,密被白色绵毛。叶对生,无柄戟形或剑形,基部半抱茎,先端渐尖,边缘微背卷,中脉凹陷,明显,全叶被毛。花有柄,单生或对生于枝顶和叶腋,萼片5,披针形,先端尖,基部截形;花瓣5,先端锐尖,白色,有多条天蓝色纵纹;柱头长,花药蓝色。花苞状如笔头,有淡蓝色斜旋纵沟。生于海拔500~2 000米的山坡草地,马、牛、羊均采食,属于中等牧草,分布在甘肃、陕西和华南、华中、西南等地。

A perennial herbaceous plant, commonly found in wet lands, wastelands, grasslands. It is distributed in southwest, east and central Loess plateau as well as Gansu and Shaanxi. It is a moderate forage, horses, cattle and sheep will graze it.

213-1 獐牙菜　　213-2 獐牙菜花序

214 秦艽

学名　*Gentiana macrophylla* Pall.
英名　Largeleaf Gentian

别名大叶龙胆、萝卜艽。多年生草本,种子繁殖。主根肥大,长圆锥形。茎直立或斜升,基部有纤维状的残存叶柄。基生叶披针形,呈莲座状;茎叶对生,基部联合,叶片披针形或矩圆状披针形,全缘。聚伞花序,蔟生于茎顶,呈头状或腋生成轮状;花萼膜质,萼齿小;花冠筒状钟形,蓝紫色。蒴果矩圆形,黄褐色。生长于亚高山草甸或疏林草地,主要分布在黄土高原及其与青藏高原接壤的地方。叶量大,草质好,牛、羊喜食马不多食。

214-1 秦艽
214-2 秦艽花序

A perennial herbaceous plant reproduced from seeds. It occurs in wastelands, forest fringes and canal banks. It is distributed in south and central Loess plateau Qinghai-Xizang plateau. Cattle, sheep and goats will graze it before flowering.

三十六 夹竹桃科

215 大叶罗布麻

学名 Poacynum hendersonii L.
英名 Largeleaf Poacynum

别名野麻、大花罗布麻。多年生。半灌木。茎直立，高1~2米，具乳汁。枝条光滑无毛。叶互生，叶片椭圆形或卵状椭圆形，纸质。花萼5裂，花冠下垂，外面粉红色，里面稍带紫色，宽钟状。雄蕊5枚，花药箭头状。花盘肉质，环状。蓇葖果双生，倒垂，种子有白色绢毛。主要分布在甘肃、青海、新疆一带，能适应干旱炎热的气候条件。多生于暖温带荒漠草原和盐渍化草甸。幼嫩枝条山羊、绵羊、骆驼均采食，枯黄后仅有骆驼采食落叶。

A perennial herb, spread by seeds and by crown buds. It occurs around desert and saling-alkali grasslands, distributed in Gansu, Shaanxi, Jilin and Xinjiang. It is excellent forage, various domestic animals like to graze, special sheep and cattle.

215 大叶罗布麻

三十七 萝藦科

216 鹅绒藤

学名 *Cynanchum chinense* R.Br.
英名 China Mosquitotrap

多年生草本。茎缠绕，通体被短柔毛。叶对生，具长柄。叶片宽三角状心形，叶面深绿色，叶背苍白色。两歧聚伞花序腋生，有花多数，花冠白色，裂片长圆状披针形。副花冠杯状，先端裂成10个丝状体，分为两轮，外轮与花冠片等长，内轮略短。蓇葖果双生或仅有一个发育，细圆柱形。种子卵状长圆形，扁平，先端具白绢质种毛。多生于灌丛、林缘、崖边、地埂，低等牧草，通常在青鲜时家畜不采食。

It is perennial creeping herb, spreading by crown buds and by seeds. It is commonly found in roadsides, among shrubs and the edge of cliffs. When it is fresh, animals do not like to graze. Cattle, sheep and goats will eat in winter.

216-1 鹅绒藤花果　　216-2 鹅绒藤

三十七 萝藦科

217 地梢瓜

学名　*Cynanchum thesioides*（Freyn.）K.Schum
英名　Bastardtoadflaxlike Swallowwort

217-2　地梢瓜花序

217-1　地梢瓜

别名女青、砂引草。多年生草本。茎直立或斜升，多从基部分枝，枝、叶和果实中含白色乳汁。叶对生，具短柄。叶片条形，全缘，中脉在叶背凸出，在叶面凹陷。聚伞花序腋生，花小，花萼5深裂，外面被柔毛。花冠白色，辐状。蓇葖果通常单生，纺锤形，中部膨大。种子多数，扁平，倒卵形，褐色，先端有白色的种毛。我国"三北"地区都有少量零星分布，生于干草原、丘陵坡地，或地埂、路边。幼嫩时牛、羊等家畜少食，干燥后喜食，骆驼常年喜食。

A semishrub, spread by rhizome and by seeds. It is commonly found in hill-slopes, sandy-soil, wastelands and roadsides, distributed throughout the Loess Plateau. Camels, sheep and goats like to graze it when green fresh.

218 杠柳

学名　*Periploca sepium* Bge.
英名　China Silkvine

灌木。全株无毛。叶对生，具柄。叶片卵状长圆形，先端渐尖，基部楔形，叶面光滑，全缘。聚伞花序腋生，5花萼深裂，裂片卵圆形。花冠紫红色，5裂片长圆状披针形，中间加厚，反折，内面具疏柔毛。副花冠杯状，10裂，其中5裂丝状伸长，先端向内弯。蓇葖果双生，角状，种子先端有白色绢质种毛。家畜不喜食，有时牛、羊在冬春枯草季采食凋落的干燥叶子。

218-1　杠柳

218-2　杠柳果穗

A shrub, commonly found in dry slopes, roadsides and forest edges. Usually, the livestock animals are not grazing this plant in green. Sometimes sheep would graze its dry leaves in winter.

三十八　旋花科

219　打碗花

219-1　打碗花　　　　219-2　打碗花花序

学名　*Calystegia hederacea* Will.
英名　Ivy Glorybind

多年生蔓生草本。根发达，横走地下，新根白色，质脆易断，老根暗黄色，易成片丛生。茎多自基部分枝，平卧或相互缠绕或缠绕他物而上升，光滑无毛。叶互生，有长柄。基部叶片长圆状心形，全缘。上部叶片三角状戟形，侧裂片平展，通常2裂，中裂片卵状三角形。花单生于叶腋，苞片2，宽卵形，包被花萼。萼片5，长圆形。花冠漏斗状，粉红色，有深色辐射纹。蒴果卵圆形，种子倒卵形。多生长在较松软的农田和果园之中，根、茎、叶均为多种家畜、家禽所喜食。喂猪的传统牧草。

A perennial viny herb, it spreading by crown buds and by seeds. It is commonly found in wastelands and farmlands, distributed throughout the Loess Plateau. The feeding value is very high and pigs and all livestock animals like it.

220　田旋花

学名　*Convolvulus arvensis* L.
英名　Field Bindweed European Glorybind

多年生蔓生草本。茎可缠绕它物上升，上部有疏柔毛。叶互生，有柄。叶形多变，基部叶多为戟形或箭形。花腋生，1~3朵。花梗上有2个狭小的苞片，远离花萼。花冠漏斗状，粉红色，漏斗边缘色深，底部色淡。蒴果球形或圆锥状。种子卵球形，黑褐色。广泛分布在农田、果园、荒山、荒沟、地埂、路边等处。优质牧草，各种家畜皆喜食。

A perennial viny herb, spread by crown buds and by seeds. It is commonly found in farmlands and wastelands, distributed throughout the Loess Plateau. It is an excellent forage, the feeding value is quite high. It is palatable for cattle, sheep rabbits and pigs.

220-2　田旋花花序　　　220-1　田旋花

221 银灰旋花

学名　*Convolvulus ammannii*
英名　Sivery-grey Glorybind

别名阿氏旋花。多年生矮小草本。通体密被银灰色的绢毛,茎高只有10厘米,平卧或斜升,多分枝。叶互生,条形或狭披针形,无柄。花小,单生于枝端,5萼片,卵圆形。花冠漏斗形,白色,带紫色条纹。蒴果球形,2裂,种子卵圆形,淡红褐色。典型的旱生植物,是荒漠草原、典型草原、草甸草原的伴生种,主要分布在"三北"地区及青藏高原。由于植株矮小,对放牧大家畜食用价值不高,青鲜时山羊、绵羊喜食。

A perennial herb, it is typical arid plant. It occurs on steppes, desert, hill slopes, distributed in north, northwest and northeast China. The palatability is good for goats and sheep, and horses and cattle do not graze.

221-1　银灰旋花

221-2　银灰旋花花序

222 藤长苗

学名　*Calystegia pellita* (Ledeb.) G.Don
英名　Densepubescent Glorybind

别名脱毛天剑。多年生草木。根细长。茎缠绕或下部直立,圆柱形,有细棱,密被灰白色或黄褐色长柔毛,有时毛较少。叶长圆形或长圆状线形,顶端钝圆或锐尖,具小短尖头,基部圆形、截形或微呈戟形,全缘;两面被柔毛,通常背面沿中脉密被长柔毛,有时两面毛较少,叶脉在背面稍突起;叶柄长,被毛同茎。花腋生,单一,花梗短于叶,密被柔毛;苞片卵形,顶端钝,具小短尖头,外面密被褐黄色短柔毛,具有如叶脉的中脉和侧脉;萼片近等长,长圆状卵形,上部具黄褐色

222-1　藤长苗

缘毛;花冠淡红色,漏斗状。蒴果近球形,种子卵圆形,无毛。多生于农田或弃耕地,分布在"三北"及陕、甘、宁、川地区。草质柔嫩,适口性好,羊、猪、兔均喜食。

A perennial herb, reproduced by seeds and by radical bud. It exists in farmlands or wastelands. It distributed in north, northwest, northeast and central China. It is a very good forage, sheep, goats, pigs and rabbits are willing to graze.

222-2　藤长苗花序

223 圆叶牵牛

学名　*Pharbitis purpurea*（L.）Voigt
英名　Roundleaf Pharbitis

一年生草本。通体被毛。茎细长缠绕，多分枝。叶互生，具长柄。叶片心形，全缘，先端尖或钝。总花梗与叶柄近等长，有单花1~5朵。花萼5片，卵状披针形，先端钝尖，基部有粗硬毛。花冠漏斗状，红色、蓝紫色或近白色，先端5浅裂。蒴果球形，种子倒卵形，黑色或暗褐色，表面粗糙。多为人工种植，也有逸生者分布在种植地附近的地埂、路旁，家畜喜食其叶和嫩茎。

An annual herb reproduced from seeds.It grows voluntarily in wastelands or farmlands and is cultivated everywhere around China.The feeding value is low but sheep,cattle and other livestock like to graze dry leaves.

223-1　圆叶牵牛

223-2　圆叶牵牛花序

224 中国菟丝子

学名　*Cuscuta chinesis* Lam
英名　Chinese Dodder

一年生寄生草本。茎细弱，有选择性的缠绕于其他草本植物体上，黄色或浅黄色，无叶。花多数，簇生，有时2个并生。花萼杯状，5裂，裂片卵圆形。花冠白色，壶状或钟状，5裂片向外反曲，将成熟的果实全部包被。蒴果近球形，略扁，种子椭圆形，淡黄褐色或褐色，表面粗糙，有白霜状突起物。黄土高原分布普遍，寄生于多种草本植物体上，以豆科、藜科植物多见，特别是紫花苜蓿上。适口性较好，牛、羊等放牧家畜都采食。

224-1　中国菟丝子

224-2　中国菟丝子花序

An annual parasitic herb reproduced from seeds.It usually parasitizes on legumes,crops or forages.It is distributed throughout the Loess Plateau as well as northeast China and many other provinces such as Shandong,Sichuan,Henan,Guangdong and Jiangsu,etc.The feeding value is moderate,some animals can graze it.

225 日本菟丝子

学名　*Cuscuta japonica* choisy
英名　dodder

225-2　日本菟丝子花序

225-1　日本菟丝子

别名豆寄生、无根草、金灯藤。一年生草本。缠绕茎,较粗壮,稍肉质,橘红色,常带紫红色瘤状斑点。茎的任何接触其他植物的部位,都可形成小的突起,这小的突起可演化成寄生根,扎入寄主植物的茎、叶柄以及叶中,从寄主的植物体内吸取水分、无机盐及各种营养物质。花成穗状花序,基部常多分枝;苞片和小苞片鳞状,卵圆形,顶端尖。花萼碗状,肉质,裂片5,背面常被紫红色的瘤状突起;花冠钟状,淡红色或绿白色,顶端5浅裂;雄蕊5枚着生于花冠裂片之间;子房球形,2室,花柱细长,柱头2裂。蒴果,卵圆形;种子褐色,表面光滑。我国南北均有分布,多寄生于树木和杂草上。牛、羊采食。

An annual parasitic herb reproduced from seeds, it parasitized on various woods. It is distributed throughout the country. Sheep and goats will graze when it is fresh.

三十九　紫草科

226　聚 合 草

学名　***Symphytum peregrinum*** Ledeb
英名　**Medicinal Collectivegrass**

多年生草本。丛生型,全株被糙毛。根肉质,粗壮发达,根皮淡红褐色,根肉白色,质脆具黏液,根冠处膨大。花茎通常单生,直立,上部分枝,在茎棱上具钩刺和窄翅,基生叶密集成莲座状,叶粗糙,背面叶脉隆起呈网状。下部叶片长椭圆形或长卵状披针形,较大,先端渐尖,基部楔形或近圆形,边缘波状,全缘。中上部叶较小,披针形或卵状披针形,无柄或近无柄,基部明显下延。聚伞花序组成圆锥状,花萼5深裂,裂片披针形,尖锐,宿存。花冠筒状钟形,5浅裂,孕蕾期紫色或紫红色,以后逐渐变为紫蓝色。小坚果卵形,黑褐色。打浆饲喂效果好。直接饲喂因通体被刚毛而适口性差,家畜不喜食。

A perennial herb that it is commonly spread by root buds. It grows quickly with a large amount of setaceous leaves. The feeding value is low and its grazing use is limited. But it can be use as pig feed.

226-2　聚合草　　　226-1　聚合草花序

227　附 地 菜

学名　*Trigonotis peduncularis*
　　　　（Trev.）Bennth.
英名　Pedunculate Trigonotis

别名鸡肠草。越年生或一年生草本。茎自基部分枝，下匍匐，上斜升，有短糙毛。叶互生，有短柄或近无柄。叶片椭圆形或椭圆状卵形，两面均有短糙毛，全缘。花序顶生，先端通常呈卷尾状。花萼5深裂，裂片先端尖锐。花冠淡蓝色，小坚果三角状锥形，棱尖锐，黑色。遍布人工草地、果园、"二阴"农田地。枝叶柔软，各种家畜喜食。

227-1　附地菜

227-2　附地菜花序

A biennial or annual herb reproduced from seeds.It is commonly found in hill slopes, grasslands and farmlands, distributed in Shaanxi, Gansu and part of the Loess Plateau.The feeding value is high and cattle, sheep, goats and pigs all like it.

228　鹤　虱

学名　*Lappula myosotis*
英名　European Craneknee

一年生或越年生草本。茎直立，分枝，上部分枝较多且长，有糙毛。叶互生，基部叶有柄，上部叶无柄。叶片条形或倒披针形，密被糙毛，先端钝，基部狭，叶缘略带波状。顶生花序，花生于苞腋的外侧，有短梗。花萼5深裂，宿存。花瓣5裂，淡蓝色，稍长于花萼。小坚果4，卵形，褐色，具疣状突起物，边缘有2～3行不等长的锚状刺。多生于砂壤土。家畜喜食，种子成熟时适口性下降，而且种子黏结绵羊毛，造成毛质受损。

An annual or biennial herb reproduced from seeds.It grows in sandy soil, occurs in farmlands and roadsides.It is distributed in south and southeast Loess Plateau.The

228-2　鹤虱

feeding value is high.Sheep, goats and cattle like to graze when it is fresh in spring.

228-1　鹤虱花序

三十九 紫草科

229 狼紫草

学名　*Lycopsis orientalis* L.
英名　Oriental ablfgromwell

229-1　狼紫草

越年生或一年生草本，种子繁殖。茎自基部分枝，斜升或直立，通体密布开展的长硬毛。叶互生，有柄或无柄；叶片匙形、倒披针形或条状长圆形，边缘皱波状。花序常呈尾卷状，苞片狭卵形或条状披针形；花萼5，花冠筒状5裂片，蓝色，小坚果，近卵形。生长在水分较好的草地、荒地或农田，分布于黄土高原东南部。幼嫩时家畜微嗜，伏糙毛影响家畜的适口性。

229-2　狼紫草花序

An annual or biennial herbaceous plant reproduced from seeds. It occurs in farmlands of hills and mountain areas, grows among hassocks along roadsides. It is distributed in north and northwest China as well as the western part of Henan. Animals will graze it.

230 倒提壶

学名　*Cynoglossum amabile* Stapf. et Drumm.
英名　China Houndstongue

一年生草本，种子繁殖。茎直立，上部多分枝，密生贴伏的短柔毛。基生叶有长柄，长圆形、长圆状披针形或椭圆形，全缘，两面密生短柔毛；茎中部以上叶无柄，长圆形或披针形。花序分枝开展，无苞片；花梗

230-2　倒提壶花序

结果时增长；花萼5深裂，花冠蓝色，5裂，小坚果4，卵形，密生锚状刺。分布于黄土高原与青藏高原接壤处的高山草地或林缘。青鲜时牛、羊采食，而不喜食，干燥后适口性更差。

An annual herbaceous plant reproduced from seeds. It occurs in mountain grassy slopes, alpine meadows and pinewoods edges. It is distributed in the southern part of Gansu, the eastern part of Tibet, Yunnan and the western part of Guizhou. Livestock will graze it before seeds are matured.

230-1　倒提壶

231 微孔草

学名　*Microula sikkimensis*（Clarke）Hemsl.
英名　Sikkim Microula

别名兰花花、紫草。一年生草本。茎斜升，被毛，高20厘米左右，自下部分枝。叶片披针形，先端渐尖，叶背和叶表具粗毛，基生叶和下部叶具柄，上部叶无叶柄，叶片自下而上逐渐变小。花序短而密集；苞片条状披针形，有短花梗；花萼具毛，深5裂；花冠天蓝色，鲜艳，喉部有5个附属物；雄蕊5，内藏；子房4裂。种子锥形，三边有棱，深褐色，有不规则的突起，种脐明显下陷成条状孔。主要分布于高海拔的青藏高原及毗邻地区。

231-1　微孔草　　　　231-2　微孔草花序

An annual herb, reproduced from seeds. It can grow at high altitudes and resist cold condition. It is distributed in southern Gansu, northwestern Sichuan, Qinghai and Tibet, etc.

四十　马鞭草科

232　马鞭草

学名　*Verbena officinalis* L.
英名　European Verbena

多年生草本。茎直立或倾斜，分枝开张，四棱形，幼时有短柔毛。叶对生，有柄或无柄。叶片卵圆形或长圆形，基生叶边缘常有缺刻或粗锯齿；茎生叶多数，3深裂或羽状深裂，裂片边缘有不整齐的锯齿，两面均有粗毛。穗状花序顶生或腋生，果实成熟时伸长。花萼与苞片近等长，花冠筒状，淡紫色或蓝色。蒴果在成熟时裂为4个小坚果，小坚果长圆形有棱。遍布我国各地。嫩枝叶家畜喜食。

A perennial herb reproduced from seeds. It is commonly found in stream sides, farmlands and wastelands, distributed over south and central Loess Plateau, the mostly in southern China. The feeding value is low and cattle would graze when fresh and sheep grazing in autumn.

232-2　马鞭草花序　　　232-1　马鞭草

四十一 唇形科

233 香薷

学名 *Elsholtzia ciliata* Hyland
英名 Common Elsholtzia

一年生草本。有特殊香味。茎直立,上部分枝,四棱形,有倒向疏柔毛。叶对生,有柄。叶片椭圆状披针形或卵形,边缘具钝齿,两面均有毛,叶背密生橙色腺点。轮伞花序多花,组成偏向一侧顶生的假穗状花序。苞片宽卵圆形,先端针芒状。花萼钟状,花冠淡紫色,先端略成唇形,上唇直立,微凹,下唇3裂。小坚果长圆形或倒卵形,黄褐色。我国大西北海拔1 000～2 000米山坡地区分布较多,猪喜食,其他家畜可食。

233-1 香薷　　　233-2 香薷花序

An annual herb reproduced from seeds. It is commonly found in hill slopes, river banks and farmlands. It is distributed all over the country except Xinjiang and Qinghai. The whole plant can be used as medicine.

234 密花香薷

学名 *Elsholtzia densa* Benth.
英名 Denseflower Elsholtzia

一年生草本,种子繁殖。密生须根。茎直立,自基部多分枝,分枝细长,茎及枝均四棱形,具槽,被短柔毛。叶长圆状披针形至椭圆形,先端急尖或微钝,基部宽楔形或近圆形,边缘在基部以上具锯齿,草质,上面绿色下面较淡,两面被短柔毛。穗状花序长圆形或近圆形,密被紫色串珠状长柔毛,由密集的轮伞花序组成。花萼钟状,果时花萼膨大,近球形,外面极密被串珠状紫色长柔毛。花冠小,淡紫色。小坚果卵珠形,暗褐色。高山草甸。秋季经霜后,家畜采食。

234-1 密花香薷　　　234-2 密花香薷花序

An annual herb, growing in wastelands and grasslands, distributed throughout the Loess Plateau. It has specially smell, the livestock do not graze.

235 水棘针

学名 *Amethystea caerulea* L.
英名 Skyblue Amethystea

别名山油子、土荆芥。一年生草本。多分枝;茎四棱形,紫色、灰紫黑色或紫绿色,被疏柔毛或微柔毛。叶纸质或近膜质,三角形或近卵形,3深裂,罕不裂或5裂;裂片披针形,边缘具粗锯齿或重锯齿,无柄,基部不对称,下延,无柄或近无柄,叶面绿色或紫绿色,被疏微柔毛或几无毛,背面无毛;叶柄紫色或紫绿色,有沟,具狭翅,被疏长硬毛。松散的二歧腋生聚伞花序,复组成总状圆锥花序,被疏腺毛;萼钟状,具10脉,其中5脉高起,外面被乳头状突起及腺毛;齿三角形,渐尖;花冠蓝色或紫蓝色,花冠管内藏或略长于花萼;檐部二唇形,外面被腺毛,上唇2裂,裂片与下唇侧裂片同形,为长圆状卵形或卵形,下唇3裂,中裂片扇形;能育雄蕊2,着生于下唇中裂片近基部,向后伸长,自上唇裂片间伸出,退化雄蕊2,着生于上唇裂片下,花冠筒中部,线形或几无;花丝略长于雄蕊。小坚果倒卵状三棱形,背部具网状皱纹,合生面占腹面达2/3以上。多生于河旁、溪边或农田,主要分布在陕、甘、川、鄂、豫、皖等省区。干燥后,牛、羊喜食。

235-2 水棘针花序   235-1 水棘针

An annual herb reproduced from seeds.It is commonly found in farmlands,riverbanks,stream sides and roadsides,distributed in north northwest east and central China.The feeding value is low and livestock animals graze a little.

236 夏至草

学名 *Lagopsis supina* (Steph.)IK.-Gal.
英名 Whiteflower Lagopsis

多年生草本。茎四棱形,常铺散,多从基部分枝。基生叶圆形或肾形,边缘具粗锯齿。轮伞花序腋生,花冠白色,二唇形,内有紫色条纹,小坚果褐色,倒卵状圆形。全国农田、果园、牧草地均有,以紫花苜蓿地最为多见。适口性差,家畜可食但不喜食。

236-3 夏至草花序

A perennial herb reproduced from seeds.It is commonly found in wastelands,it is serious weeds in rape-seeds and alfalfa.It is distributed throughout the country.The seeding value is quite low,and most animals do not like to graze when it is fresh.

236-2 夏至草幼苗   236-1 夏至草

237 益母草

学名 *Leonurus heterophyllus* Sweet
英名 Wormwoodlike Motherwort

越年生草本，种子繁殖。茎直立，多在中部以上分枝，钝四棱形，有毛。叶对生，有柄或近于无柄。茎下部叶轮廓卵形，掌状3裂，其上再分裂，中部叶常3裂成长圆形裂片，花序上的叶呈条形或条状披针形，全缘或具疏齿。轮伞花序，小苞片刺状，花萼筒状钟形，花冠粉红、淡紫红色。小坚果长圆状三棱形，褐色。我国各地多有分布，主要生长在荒地、河滩、田边、路旁或山沟草丛中。家畜不喜食，且含有leonurin结晶物质，有些家畜食后会引起中毒。

A biennial herb reproduced from seeds.It is commonly found in waste-lands,river beaches and hassocks in the mountain valleys.It is distributed all over the country. It can be used as forage.Cattle,sheep and pigs will graze when it is fresh in early spring.

237 益母草

238 细叶益母草

学名 *Leonurus sibiricus*
英名 Siberia Motherwort

一年生或二年生直立草本。有短而贴生的糙伏毛。茎中部叶轮廓卵形，掌状3全裂，裂片再分裂成条状小裂片，花序上的叶明显3全裂，中裂片复3裂，全部小裂片均条形。轮伞花序轮廓圆形，下有刺状苞片；花萼筒状钟形，5脉，齿5，前2齿靠合；花冠粉红至紫红，花冠筒内有毛环，檐部二唇形，下唇短于上唇1/4，上唇外密被长柔毛，下唇3裂，中裂片倒心形。小坚果矩圆状三棱形。幼嫩时可作牛、羊饲料。

A biennial or annual herb reproduced from seeds.It grows in farmlands,wastelands and hassocks in the mountain valleys.It is distributed in south and central Loess plateau.The feeding value is low.

238-2 细叶益母草花序

238-1 细叶益母草

239 紫 苏

学名　*Perilla frutescens* Britt
英名　Common Perilla

一年生草本,有香味。茎四棱,紫色或绿色,被倒生毛。叶对生,宽卵形或卵圆形,边缘具粗锯齿,表面绿色,背面紫色,或两面均为绿色。轮伞花序常组成顶生或腋生的总状花序,生于一侧,花小,二唇形,红色或淡红色。小坚果近球形,具网纹。黄土高原不少地方有栽培种,家畜不乐食。

An annual aromatic herb reproduced from seeds.It grows voluntarily or cultivated in south and central Loess Plateau.When it is fresh,animals do not graze,but will consume this plant as hay or powder.

239-2　紫苏花序　　　　239-1　紫苏

240 百里香

学名　*Thymus mongolicus* Ronn
英名　Mongo Thyme

别名地椒。多年生小灌木。茎多分枝,匍匐或斜升。叶对生,椭圆形,先端钝尖,基部楔形,全缘,下面有腺点,具短柄。轮伞花序紧密,排列成头状,花萼狭钟形;唇形花冠,紫红色或粉红色。小坚果近圆形,光滑,暗褐色。百里香分布于典型的草原地带。幼嫩时小家畜及羊、马喜食,花期前后家畜不食,秋季枯黄后绵羊、山羊均喜食。

A small shrub,commonly found on sunny hill slopes,field-sides and roadsides.It is distributed all over the Loess plateau as well as Qinghai, Hebei and Inner Mongolia.The feeding value is moderate,cattle,sheep and goats will graze it.

240-2　百里香花序　　　　240-1　百里香

241 冬青兔唇花

学名 *Lagochilus ilicifolius* Bunge
英名 Hollyleaf Lagochilus

别名冬青叶兔唇花。多年生草本。茎基部木质化，多分枝，密被短柔毛，混生疏长柔毛。叶革质，楔状菱形，先端具5~8齿裂，齿端具芒状刺尖，基部楔形，两面无毛；轮伞花序具2~4个，着生于上部叶腋；花冠淡黄色，密生短柔毛；小坚果窄倒三角形，顶端截平。典型的旱生植物，荒漠草原中常见。骆驼和羊春、夏、秋皆采食，马夏、秋少量采食，牛不食。

A perennial herbaceous plant, a typical drought-enduring plant, occurring in desert grasslands and hill-slopes. It is distributed in north west Loess plateau. Sheep and camel will graze it in spring and summer.

241-1 冬青兔唇花　　241-2 冬青兔唇花花序

242 甘西鼠尾草

学名 *Salvia przewalskii* Maxim
英名 Przewalsk Sage

别名甘肃丹参。多年生草本。我国特有植物。根木质，直伸，圆柱状，外皮红褐色。茎自基部分枝，上升，丛生，上部节上分枝，密被短柔毛。基生叶具长柄，茎生叶对生，均密被微柔毛；叶片三角状或椭圆状戟形，稀心状卵圆形。轮伞花序2~4个，疏离，组成顶生的总状花序，或再组成圆锥状花序；苞片卵圆形或椭圆形，全缘，两面被毛；花萼钟状，外面密被腺毛，并杂有红褐色腺点，二唇形；花冠紫红色，散布红褐色腺点，小坚果倒卵形。分布于青藏高原及毗邻的黄土高原西南缘交错地带。

A perennial herbaceous plant, commonly found in mountain grasslands, alpine meadows and hill slopes. It is distributed in the part of Qinghai-Xizang plateau and the southwest of Loess plateau.

242-1 甘西鼠尾草花序

242-2 甘西鼠尾草

243 鼬瓣花

学名 *Galeopsis bifida* Boenn
英名 Bifid Hempnettle

243-2 鼬瓣花花序

243-1 鼬瓣花

别名野芝麻、野苏子。一年生草本。茎直立或斜升,钝四棱形,在节上加粗但在干时则明显收缢,节上密被多节长刚毛,节间混生下向具节长刚毛及贴生的短柔毛,有时杂以腺毛。茎叶卵圆状披针形或披针形,先端急尖或渐尖,基部渐狭至宽楔形,边缘有锯齿,上面贴生具节刚毛,背面疏生微柔毛,间夹有腺点;叶柄长,被短柔毛。轮伞花序含多花,苞片线形至披针形,有刺毛;花冠白色、黄色或粉紫红色,冠管漏斗状,冠檐上唇卵圆形,先端钝,具不等的数齿;子房无毛,褐色。小坚果倒卵状三角形,褐色有秕鳞。

An annual herb, reproduces from seeds. It exists in roadsides, farmlands, forest fringes and grasslands. It is distributed in north and northwest China as well as the southwest regions of China. Animals grazing are limited when mature.

244 糙苏

学名 *Phlomis pratensis* L.
英名 Meadow Jerusalemsage

多年生草本。茎直立,少分枝,疏被绵毛,略显四棱形。叶对生,叶片呈心脏形或长圆形,先端钝,基部心形,边缘有锯齿。下部和基生叶大而柄长,向上叶片逐渐变小,叶柄变短,最上部苞叶无柄。轮伞花序腋生,无花梗,着生于茎枝上部,花萼筒状,裂齿5,辐射对称。花冠唇形,粉红色,上唇弯曲,下唇3裂。坚果小,卵形无毛。主要生长在海拔1 500～2 500米的山坡草地,喜平缓阳坡。属于高山地区的优良牧草,牛、羊喜食,开花结实后,采食减少。因其茎高叶大,也是刈割调制干草的主要对象。

A perennial herb, reproduced from seeds. It is commonly found in fertility farmlands or wastelands, distributed in south and central Loess Plateau and south of the China. The feeding value is low. Cattle and sheep graze a little when it is dry.

244-1 糙苏

244-2 糙苏花序

四十一　唇形科

245　香青兰

学名　*Dracocephalum moldavica* L.
英名　Fragrant Greenorchid

245-1　香青兰

245-2　香青兰花序

一年生草本，种子繁殖。有特殊气味。茎直立，中下部分枝，略显四棱形，有毛。叶对生，具柄；叶片卵状三角形、披针形或条状披针形，叶缘具三角状牙齿或疏锯齿，叶基部的齿常有长刺。轮伞花序生于茎或分枝上部；花萼二唇形，上3下2。花冠淡蓝紫色，小坚果三棱状长圆形。生于山坡草地、林缘或河滩沙地，分布于"三北"地区。

An annual herbaceous plant reproduced from seeds.It occurs in forest fringes of mountain area,grassy hill-slopes,farmlands and sandy lands of river beaches.It is distributed throughout the country.The feeding value is moderate.

246　丹　参

学名　*Salvia miltiorrhiza* Bunge
英名　Dan-shen

246-2　丹参花序

246-1　丹参

别名血生。多年生草本，种子和根芽均可繁殖。根肉质、肥大、朱红色、圆柱形。茎直立，四棱形，被黄白色柔毛及腺毛。叶对生，奇数羽状复叶，小叶3~5片，卵形至宽卵形。轮伞花序组成顶生或腋生的总状花序，花蓝紫色；苞片披针形，被绿毛，花萼钟状；小坚果椭圆形，黑色。分布于黄土高原、华北、华中。低等牧草，舍饲家畜少量采食干燥或青鲜的叶片及枝条。

A perennial herb,with big succulent roots,it grows under forest or roadsides.It is distributed in Shanxi,Shaanxi,Jiangsu and Fujian provinces.The feeding value is low,livestock refuse to graze when fresh,but like to graze as hay.

247 薄 荷

学名　*Mentha haplocalyx* Briq
英名　Mint

247-2 薄荷花序

多年生草本,根茎和种子繁殖。茎叶揉之有薄荷的特殊气味。匍匐根茎较粗壮,节处生根。地上茎直立或略倾斜,有分枝,四棱形,有毛。叶对生,具柄或近于无柄;叶片长圆状披针形至披针状椭圆形,边缘具疏齿,两面均有毛。轮伞花序腋生,球形,有梗或无梗;花萼筒状钟形;花冠淡紫色、淡红色或白色,裂片4。小坚果卵球形,黄灰色或栗褐色。常生于河边、渠边、水湿地,全国大部分地区均有。鲜草家畜不食,与其他牧草混合调制成干草或草粉牛、羊采食。

A perennial herbaceous plant, spread by rhizome and by seeds.It occurs in wet lands of water sides, hassocks in the valleys, canals and farmlands, and is distributed all over the country.The feeding value is low with animals will graze as dry.

247-1 薄荷

248 并头黄芩

学名　*Scutellaria scordifolia* Fisch.ex Schrank.
英名　Skullcap

多年生草本。多直立,根粗壮而肉质。茎四棱形,叶对生。叶片披针形至线状披针形。总状花序顶生,花萼在上唇的背面有一圆形的盾片,

248-2 并头黄芩花序

果时增大,花冠紫色或蓝色,二唇形,基部常膨大而膝曲。小坚果黑褐色,具小瘤。多生于600～1500米的丘陵或山坡草地,是有些地方草地主要的组成草之一。适口性不佳,家畜只采食嫩枝叶。

A perennial herbaceous plant reproduced from seeds.It is commonly found in mountain grassy slopes, wastelands and hassocks in the valleys.It is distributed in northeast, southwest China and Gansu as well as Shanxi.The feeding value is low.

248-1 并头黄芩

249　黄　芩

学名　*Radix Scutellariae*
英名　Baikal Skul lcapRoot

别名黄金茶、山茶根。多年生草本。根粗壮而肉质。茎丛生，四棱，具细条纹，近无毛或被上曲至开展的微柔毛。叶对生，披针形至条状披针形，全缘，下面密被下陷的腺点。总状花序顶生，花偏生于花序一侧；花萼二唇形，果时增大；花冠紫色、紫红色至蓝紫色，花冠筒近基部明显膝曲；雄蕊4，二强。小坚果卵球形，黑褐色，具瘤。多生于600~1 500米的丘陵草地，是黄土高原南部临近秦岭山脉有些草地的主要草类之一，低等牧草，牛、羊采食其鲜茎。

A perennial erect herb, commonly found in hilly slopes and grasslands.It is distributed in provinces in north, northeast and southwest China as well as Shaanxi and Gansu.The feeding value is low and livestock would graze green branches.

249-1　黄芩　　　249-2　黄芩花序

四十二　茄　科

250　枸　杞

学名　*Lycium chinense* Mill.
英名　Chinese Wolfberry

多年生灌木。具短刺。茎直立，分枝稀疏，表皮灰白色，枝条下部坚韧，上部往往呈下垂状。叶互生，卵形或卵状披针形，先端圆钝，基部渐狭，全缘，两面均无毛。花单生或簇生于叶腋，有长梗。花冠淡紫色，5裂，边缘有毛；花萼钟状，3~5裂。浆果卵形或长椭圆状卵形，成熟时红色。主要分布在西北较干旱的地区，也有人工种植药用的。幼嫩枝叶家畜喜食，老枝具刺，所以作为饲草利用率低。

A shrub with thorns, spread by crown buds and by seeds.It is a common plant found in roadsides, wastelands and field ridges and distributed over all parts of the Loess Plateau.The nutritive value is high.Its leaves are palatable for grazing livestock.

250-2　枸杞花序　　　250-1　枸杞

251　黑果枸杞

学名　*Lycium ruthenicum* Murray
英名　Blackfruit Wolfberry

别名苏枸杞。落叶灌木。多分枝，枝条坚硬，常呈"之"字形弯曲，白色，枝上和顶端具棘刺。叶2～6片簇生于短枝上，肉质，无柄，条形、条状披针形或圆棒状，先端钝圆。花1～2朵生于棘刺基部两侧的短枝上，花梗细；花萼狭钟状，2～4裂；花冠漏斗状，传部较檐部裂片长2～3倍，浅紫色。雄蕊不等长。浆果球形，成熟后紫黑色，种子肾形，褐色。适生于盐化砂砾质荒漠上，我国主要分布于西北。骆驼一年四季均采食，羊仅于春季喜食嫩枝叶，属于干旱地区的优良牧草。

251-2　黑果枸杞浆果　　251-1　黑果枸杞

An echinate shrub, spreading by radical bud and by seed. It is a common plant found around sand-beaches and sandy lands in the desert. It is distributed in northwest Loess plateau as well as Xinjiang, Ningxia, Qinghai and Inner Mongolia. The leaves and tender braches are very good forage, and animals, especially camels, will graze.

252　挂金灯酸浆

学名　*Physalis alkekengi* L.
英名　Groundcherry

一年生或多年生草本。根茎长，横走。茎直立，不分枝，有纵棱，节稍膨大。单叶互生，或茎上部假对生，长卵形、宽卵形或菱状卵形，顶端渐尖，基部偏斜，全缘，波状，有粗齿，具柔毛，叶柄长。花单生于叶腋，花萼钟状，绿色，被柔毛。花冠辐状，白色，5浅裂，外面有短柔毛。浆果球形，由膨大的宿萼包被。宿萼卵形，远大于浆果，基部稍内凹，橙红色。种子多数，肾形，淡黄色。

252　挂金灯酸浆

An annual herb reproduced from seeds. It exist mainly in hill slopes under forest or gardens, distributed over all parts south to the Yangtze River and Shaanxi province, etc. The feeding value is moderate. Cattle and sheep like to graze in autumn.

253 龙 葵

学名 *Solanum nigrum* L.
英名 Black Nightshade

一年生草本。茎直立,多分枝,无毛。叶互生,具长柄。叶片卵形,全缘或有不规则的波状粗齿,两面光滑或有疏短柔毛。伞形聚伞花序,呈短蝎尾状,腋外生,有花多朵,花梗下垂。花萼杯状,5 裂。花冠白色,5 裂,辐状,裂片卵状三角形。浆果球形,成熟时黑色或紫黑色。种子近卵形,扁平。该草喜肥沃土壤,多生于农田,家畜不喜食。

An annual herb reproduced from seeds. It occurs in grasslands farmlands and nearby villages, distributed in throughout the country. It is a toxic plant, usually animals do not graze.

253-1 龙葵　　　253-2 龙葵花序

254 红果龙葵

学名 *Solanum alatum* Moench.
英名 Redfruit Dragon Mallow

一年生草本,种子繁殖。茎斜升或直立,中上部多分枝,无毛。叶互生,具柄,叶片卵形,沿中脉向内折,全缘,或边缘有波状细齿,叶色淡黄绿。聚伞花序短蝎尾状,腋外生,花多朵,梗下垂,花萼杯状,裂片 5;花冠白色。与龙葵的主要区别在于成熟时果实朱红色,较小,果皮薄;小枝有棱状狭齿。该草喜肥沃土壤,多生于路旁、荒地、农田或草地。鲜嫩时家畜多不采食。

254-1 红果龙葵

254-2 红果龙葵浆果

An annual herbaceous plant reproduced from seeds. It occurs in wastelands, roadsides and forest edges, and is distributed in Gansu, Qinghai, Shanxi Xinjiang and other provinces. Usually, livestock have limited grazing.

255 青杞

学名　*Solanum septemlobum* Bge.
英名　Sevanlobed Nightshade

255-2　青杞花序

别名野茄子。多年生草本或半灌木。茎直立，多分枝，有棱，疏生短柔毛或无毛。叶互生，具柄；叶片卵形，羽状深裂，裂片多为披针形，两面着生疏短柔毛，尤以叶脉和边缘处最密。二歧聚伞花序顶生或腋外生。花萼小，杯状，5裂，裂片三角形，宿存。花冠蓝紫色，5裂，裂片长圆形。浆果近球状椭圆形。未成熟时青绿色，成熟后变为红色。种子扁圆形，黄色。我国"三北"地区均有分布，适口性较好。

A herbaceous or semifruticose perennial, spread by crown buds and by seeds. It is commonly found in sunny hill slopes, roadsides, orchards and farmlands. It is distributed in north, northeast and northwest China. Cattle and sheep like to graze this plant, special after an autumn frost.

255-1　青杞

四十三　玄参科

256 柳穿鱼

学名　*Linaria vulgaris* Mill.
英名　Yellow Toadflax

多年生草本，根芽和种子繁殖。茎直立，通常少分枝，无毛。叶互生，无柄；叶片条形或条状披针形，全缘，无毛。总状花序顶生，各部被腺毛，少无毛；花萼5深裂，裂片披针形，宿存；花冠黄色，花冠筒由前面延伸为距，蒴果近球形，种子宽椭圆形，扁平，边缘有薄翅。常生长在山坡草地植被稀疏处，长江以北常见。低等牧草，一般情况下牛、羊不食。

256-2　柳穿鱼花序

A perennial herbaceous plant spread by crown buds and by seeds. It occurs in sandyland, grassland of hill-slopes and roadsides. It is distributed over the provinces and autonomous regions north to the Yangtze River. Animals do not graze commonly.

256-1　柳穿鱼

257 匍茎通泉草

学名 *Mazus miquelii* Makino

匍茎通泉草与通泉草相似,主要差异在于:该草除了直立茎外还有匍匐茎,有时茎带紫色,叶面中部脉上有紫

257-1 匍茎通泉草

257-2 匍茎通泉草花序

色斑,边缘具粗齿;花白色而有紫斑或带蓝紫色。

An annual herb, reproduces from seeds.It is a common plant in farmlands, wastelands and roadsides.It is distributed over Gansu, Shaanxi, Jia ngsu, Hunan and Taiwan, etc.The feeding value is moderate or high, livestock animals will graze it when fresh.

258 通泉草

学名 *Mazus japonicus* (Thunb.) O.Kuntze
英名 Japanese Mazus

一年生草本。茎自基部分枝,倾斜或直立。基生叶多数倒卵状或匙状,边缘有不规则的粗钝齿,基部下延至叶柄两侧呈翅。茎上几乎无叶,全是花序。花萼钟状,5裂,裂片与萼筒近等长。花冠淡蓝紫色或白色,二唇形,上唇2裂,下唇3裂。蒴果球形,稍露出萼外,种子长圆形。分布华北、西北。多生于小麦、豆类、油菜等作物地。草质优良,多种家畜喜食。

258-1 通泉草　　　　258-2 通泉草花序

An annual herb reproduced from seeds.It is commonly found in damp farmlands, wastelands and roadsides.It is distributed over various parts south to Hebei and Gansu provinces.The feeding value is high and livestock and poultry will graze it.

259 北水苦荬

学名　Veronica anagallis-aquatica L.
英名　Watery Speedwell

259-2　北水苦荬花序

259-1　北水苦荬

多年生草本,根茎和种子繁殖。茎直立。叶对生,无柄,叶片长圆状卵形至长圆状披针形,基部半抱茎,全缘或有细锯齿。总状花序腋生,花多数,花梗弯曲上升,与花序轴成锐角,与苞片近等长;花萼4深裂,裂片卵状披针形;花冠淡蓝色、淡紫色或白色,管部极短。蒴果卵圆形,种子卵形至椭圆形,半透明状,无光泽。多生于低湿地或浅水中,也是稻田杂草。家畜喜食。

A perennial herbaceous plant, spread by rhizome and by seeds. It occurs in shallow water, damp low land and wet farmlands. It is distributed over region north to the Yangtze River as well as northwest and southwest China. Livestock grazing is limited.

260 甘肃马先蒿

学名　Pedicularis kansuensis Maxim.
　　　subsp.kansuensis f.albiflora

260-2　甘肃马先蒿花序

260-1　甘肃马先蒿

多年生草本。高15~20厘米,干时稍变黑。根茎细长鞭状,须根丛生于根茎周围。茎直立,不分枝,被短柔毛。叶互生,近基部较密,短柄,上部渐疏;叶片线状披针形,先端钝,叶缘羽状浅裂,裂齿全缘,背卷;叶两面疏生短毛。花序短总状,上部较密。萼钟形,密生腺毛,花冠盔部紫红色,余为黄色,幼果卵形而扁。分布在临近青藏高原的地方,放牧家畜采食。

A perennial herbaceous plant, commonly found on mountain grassland and alpine meadows. It is distributed in southwest Loess plateau and Qinghai-Xizang plateau. The feeding value is moderate and cattle and sheep will graze it.

261 轮叶马先蒿

学名　*Pedicularis verticillata* L.
英名　Whorledleaf Woodbetony

261-2　轮叶马先蒿花序

261-1　轮叶马先蒿

多年生草本，种子繁殖。茎丛生，直立。基生叶矩圆形至条状披针形，羽状深裂至全裂，裂片有缺刻状齿，齿端有白色胼胝；茎生叶常4枚轮生，叶柄较短，叶片较基生叶宽短。总状花序顶生，稠密。花萼球状卵形。花冠紫红色，唇形。蒴果披针形，带褐色。种子半圆形，黑褐色。分布于黄土高原与青藏高原接壤的河谷、水沟湿地，青鲜时羊喜食，干燥后牛也乐于采食。

A perennial herbaceous plant, it occurs in mountain grassy slopes, alpine meadows and grows among hassocks or forest. It is distributed in north, northwest China as well as Sichuan and Tibet. It is a moderate forage, cattle and sheep will graze it.

262 半扭卷马先蒿

学名　*Pedicularis semitorta* Maxim
英名　Halftwisting Woodbetany

多年生草本，种子繁殖。茎直立，有明显的纵棱，无分枝，下部表皮略带紫红色。叶互生，无柄，羽状全裂，裂片呈缺刻状齿，幼嫩时稍向背后反卷。总状花序顶生，小花较大，排列松散，花冠盔部形成一半扭卷状如羊角的距，紫红色，为该草最明显的标志。分布在青藏高原及临近的地方，喜草甸草原或潮湿地带，家畜喜食叶片。

262-2　半扭卷马先蒿花序

A perennial herbaceous plant reproduced from seeds. It occurs in wet lands, mountain meadows and water sides in the gullies. It is distributed between Qinghai-Xizang plateau and Loess plateau. In spring or autumn, livestock will graze fresh leaves and branches.

262-1　半扭卷马先蒿

263 藓状马先蒿

学名 *Pedicularia muscicola*
英名 Musciform Woodbetony

多年生草本。丛生，少分枝。茎秆幼嫩质脆，半直立或斜升，密被腺毛，白色或略带紫红色。叶片轮廓卵圆形至宽披针形，羽状全裂，小裂片具不规则深裂或浅裂，淡黄绿色，秋后略带淡紫色。花多单生于枝顶，有长距，状如羊角，花冠淡紫红色。喜生于潮湿环境，放牧家畜喜食。

A perennial herb, it occurs on sunny slopes and grasslands, distributed over all parts of the country. It is a toxic plant for the animals. The livestock usually do no graze.

263-1 藓状马先蒿

263-2 藓状马先蒿花序

264 小米草

学名 *Euphrasia tatarica* Fisch.
英名 Pectinate Eyebright

一年生或二年生直立草本，种子繁殖。少分枝，散生白色微卷毛，茎四棱形，无基生叶。茎生叶对生，阔卵形或近圆形，先端钝，基部楔形；每边具2~4个钝或尖头齿，密被短腺毛。穗状花序顶生，花密集，苞片叶状；花冠2唇形，淡黄白色，上唇2裂，下唇3裂，先端稍有凹缺。蒴果长椭圆形，与宿存的花萼近等长。生于山坡草丛之中。

264-2 小米草花序

A biennial or annual herbaceous plant reproduced from seeds. It occurs in mountain grasslands, hill slopes and wastelands, and is distributed in southwest Loess plateau and Qinghai-Xizang plateau.

264-1 小米草

265 地　黄

学名　*Rehmannia glutinosa* Libosch.
英名　Adhesive Rehmannia

别名生地,狼耳朵。多年生草本。通体被白色长腺毛,肉质根肥大,茎直立。基生叶发达,丛生,有长柄。叶片倒卵状披针形或长椭圆形,边缘具不整齐的齿,叶面呈波状皱缩不平,茎生叶小而少或无。总状花序顶生,密布长腺毛,有时花从茎基部生出。花萼5齿,萼筒坛状。花冠筒状,有毛,先端5浅裂,淡紫红色或黄白色,间有紫纹。蒴果卵形,种子近卵形,有网纹。

265　地黄

野生种零星分布在低海拔的沟壑下部和路旁,幼嫩时家畜喜食,经霜枯萎后,采食性增强。

A perennial herb, spread by crown buds and by seeds. It occurs in hill slopes and creek sides, distributed over south and central Loess Plateau. The feeding value is moderate and cattle are willing to graze it when fresh or dry. Sheep and goats graze a little.

266　阿拉伯婆婆纳

学名　*Veronica persica* Poir.
英名　Iran Speedwell

越年生或一年生草本。通体被柔毛。多为丛状生长,茎自基部分枝,匍匐或斜升。基部叶对生,具短柄或无柄。叶片卵圆形或卵状长圆形,边缘有粗钝齿。顶生花序,花单生于苞腋。苞叶与茎叶同形,互生,短于花梗。花萼4裂,裂片卵状披针

266-1　阿拉伯婆婆纳

266-2　阿拉伯婆婆纳花序

形。花冠淡蓝色,有深色脉纹多条,花冠筒极短,4裂,裂片宽卵形。种子椭圆形,黄色,表面有粒状突起,腹面有凹陷。家畜较喜食。

A biennial or annual herb reproduced from seeds. It is commonly found in farmlands and grasslands, distributed in Shaanxi, Gansu and part of the Loess Plateau. Cattle, sheep and pigs like to graze before flowering and sheep will graze it after dry.

267 光药大黄花

267 光药大黄花

学名　*Cymbaria mongolica* Maxim.
英名　Mongol Cymbaria

多年生草本。茎丛生，低矮，茎基部为鳞片所覆盖，密被短柔毛。叶对生，茎上部叶近互生，无柄；叶片长圆状披针形至条状披针形，两面均有柔毛，全缘。花单生于叶腋，较大，每茎有花一至多朵。花萼先端有5~6齿裂，花冠黄色，二唇形，上唇盔盖状向外侧反卷，下唇3裂，开展。蒴果长卵状。放牧反刍家畜较喜食。

A perennial small herb.It occurs in dry hill slopes, commonly found in grasslands and gobidesert.It is mainly distributed over the Loess plateau.Camels like to graze it year around, sheep graze it in autumn only.

四十四　紫葳科

268 红花角蒿

268-2　红花角蒿花序

学名　*Incarvillea sinensis* Lam.
英名　China Hornsage

一年生草本。茎直立，圆柱形。茎基部叶对生，上部叶互生；叶片2~3回羽状全裂，小裂片条形或条状披针形。总状花序顶生，小花数量多少变幅较大；花萼钟状，萼齿5。花冠淡红色，略成二唇形。蒴果为长角果，表面有棱。种子卵形，扁平，边缘有较宽的白色膜翅。幼嫩时家畜喜食，花后粗老茎多叶少则适口性下降。

An annual herb reproduced from seeds.It grows on hills, wastelands and grasslands, distributed in Gansu, Shaanxi, Sichuan and Qinghai provinces.The feeding value is low.Cattle and sheep like to graze fresh branches and leaves.

268-1　红花角蒿果

269 黄花角蒿

学名　*Incarvillea lute* Bur. et Franch
英名　Yellowflower Hornsage

多年生直立草本。茎直立,被淡灰褐色的微柔毛,根状茎肉质。叶互生,羽状;侧生小叶椭圆形,上部小叶常不完全与叶轴分开且有第二主脉与中肋平行。花序顶生,总状;花萼钟状,常散生黑斑;花冠淡黄色或红色,有时在喉部有褐色或深红色斑点和条纹,裂片顶端圆或微凹,生有短柄腺体;雄蕊4枚。蒴果粗糙,具4棱,顶端渐尖,向前弯曲;种子卵形或圆形,平凸,黄褐色,上面常有光泽,下面有淡灰色的微柔毛。分布于黄土高原及云南、四川等省。青鲜时家畜不采食。

It is an perennial herb, reproduces from seeds.It grows on hill slopes, wastelands and grasslands, distributed in Gansu, Shaanxi, Sichuan and Qinghai provinces.The feeding value is low.Cattle and sheep like to graze fresh branches and leaves.

269-2　黄花角蒿花序

269-1　黄花角蒿

四十五　列当科

270　列　当

学名　*Orobanche caerulescens* Stuph.
英名　Skyblue broomrape

一年生寄生植物,高20~40厘米。茎不分枝,基部往往加粗,干后为黄褐色,覆盖短腺毛,被有稀疏的鳞片;鳞片卵状披针形或披针形,被腺毛。花序卵圆形或短圆柱形,花较密,少稀疏;苞片卵状披针形,比萼片稍长或近等长,沿有小苞片;萼片淡褐色,通常2裂,裂齿窄披针形;花冠筒状,蓝色或紫色,干后还往往保留颜色,向冠喉逐渐扩大,通常被有稀疏的短腺毛,上唇唇瓣宽,圆形,下唇较狭,渐尖,唇瓣的边缘具圆齿,无毛;雄蕊着生在花冠筒下部1/3处,花丝基部有短毛,上部无毛,花药沿缝线有毛;花柱近无毛。蒴果长圆状椭圆形,种子多数,小。多生于沙丘、干草原、砾石沙地和戈壁,我国东北、西北及四川、山东有分布,常寄生于菊科蒿属植物根上,影响牧草生长。

It is an annual parasitic herb, usually parasitizes on roots of wormwood.It is commonly found in drought grasslands, gobi-desert and sandy lands, distributed northeast, northwest and north China as well as Sichuan and Shandong provinces.

270　列当

四十六 车前科

271 车前

学名 *Plantago asiatica* L.
英名 Asiatic Plantain

别名车轱辘菜。多年生草本。根茎短而肥厚,不明显,簇生多数须根。叶基生,具叶柄,卵形或宽卵形,先端钝圆,基部圆形至楔形,边缘有不整齐的波状疏钝齿,两面无毛或有短柔毛。花葶数条,直立,有短柔毛。穗状花序顶生,细圆柱形。花小,多数,密集,绿白色或微带紫色。花冠膜质,4裂,裂片披针形。蒴果椭圆形,成熟时盖裂,内含5~8粒种子,种子长圆形,黑棕色。生于潮湿的水渠、路旁、河滩或弃耕地,为伴生植物,遍布我国,各种家畜乐于采食,猪尤其喜食。

A perennial herb reproduced from seeds. It is commonly found in gullies, roadsides, river beaches and fairly moist wastelands. It is distributed over various parts of the country. The feeding value is high and sheep and goats will graze before immature, pig like to graze it too.

271-2 车前根

271-1 车前

272 大车前

学名 *Plantago major* L.
英名 Rippleseed Plantain

别名车前、车前草。多年生草本。高15~30厘米,根状茎粗短,须根系,基生叶直立,纸质。叶柄长。叶片卵形或宽卵形,先端钝圆,边缘波状或有不规则的锯齿,两面疏被柔毛。花葶数条,近直立,长10~20厘米。花小,密集,淡绿色,排列成穗状花序。花冠干膜质,蒴果圆锥状,种子矩圆形。主要分布在温带地区。喜潮湿,多生于水肥较好的草地、农田、路旁、河湖水边、沟底,花葶和基生叶均柔软多汁,适口性较好,青绿期长,各种家畜均采食,绵羊、家禽最喜食。

272-1 大车前

A perennial herb, it is spread by seed, crown buds and stolons. It is commonly found in river sides and hill-slope grasslands, and distributed in Gansu, Shaanxi, Liaoning and Shandong, etc. The feeding value is high for grazing animals.

272-2 大车前根

273 小车前

学名　*Plantago minuta* Pall.
英名　Little Plantain

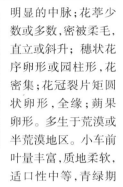

一年生草本,种子繁殖。根细长。基生叶平铺地面,叶条状或条状披针形,先端渐尖,基部渐狭,全缘,干燥后很脆。两面密被绵毛,有明显的中脉;花葶少数或多数,密被柔毛,直立或斜升;穗状花序卵形或园柱形,花密集;花冠裂片矩圆状卵形,全缘;蒴果卵形。多生于荒漠或半荒漠地区。小车前叶量丰富,质地柔软,适口性中等,青绿期各种家畜采食,羊喜食,牛、马次之。

273-1　小车前　　273-2　小车前花序　　273-3　小车前根

An annual herb reproduced from seeds. It grows on meadows, light salinity meadows, and also on roadsides, fields and nearby residential areas in the Loess Plateau. It is an excellent forage with large amount of leaves. It is palatable for sheep and goats.

274 平车前

学名　*Plantago depressa* Willd.
英名　Depressed Plantain

别名车轮菜、车串串。一年生或越年生草本。直根圆柱状,基生叶,直立或平铺地面,椭圆形、椭圆状披针形或卵状披针形,先端锐尖或稍钝,基部下延成柄,边缘有不规则的疏齿,两面被柔毛或无毛,有弧形脉5~7条。花葶直立或弧曲。穗状花序,中、上部花较密生,下部花较疏;苞片三角状卵形;萼四裂,白色,膜质。花冠淡绿色,蒴果圆锥状,成熟时盖裂,含种子4~5粒,种子矩圆形,黑棕色。

274-2　平车前花序　　274-3　平车前根　　274-1　平车前

具有旱生特点,能忍耐轻度盐渍化土壤,陕、甘、宁黄土丘陵地区分布较为广泛。中等饲用植物,马、牛、羊、驼均乐食,幼嫩时猪、兔亦喜食。

An annual or biennial herb reproduced from seeds. It grows on meadows, light salinity meadows, and also on roadsides, fields and nearby residential areas in the Loess plateau. It is an excellent forage with large amount of leaves. It is palatable for sheep and goats.

四十七　茜草科

275　茜草

学名　*Rubia cordifolia* L.
英名　India Madder

多年生草本，根系发达。茎四棱，细长且多分枝，茎棱上密布倒生小刺，依靠小刺粘贴攀缘它物而向上生长。叶四片轮生，具长柄。叶片卵形至卵状披针形，全缘，背部叶脉和叶柄常有倒生小刺。聚伞花序通常排列成大而疏松的圆锥花序，顶生或腋生。花小，黄白色，有短梗。花冠5，深裂，辐状。浆果近球形，成熟后暗红色。种子黑色，球形。无论鲜草，还是干草，家畜均可食，食性差。

A perennial herb, a common weed, spreading by seeds and by crown buds. Found in orchards and farmlands. It is distributed around the Yellow River and Yangtze River valleys. Usually the livestock don't like it, and sometimes, cattle and sheep graze a little, but rabbits like it.

275-1　茜草　　275-2　茜草花序

276　猪殃殃

学名　*Galium aparine* var. *tenerum* Cleavers
英名　Tender Catchweed Bedstraw

一年生蔓生草本。茎多自基部分枝，纤细，四棱形，棱上有倒生小刺毛。叶轮生，6~8片，条状披针形，先端凸尖，基部楔形，全缘，叶缘及背面均有倒生的小刺毛，无柄；托叶各式，在叶柄间或在叶柄内，有时与普通叶一样，宿存或脱落。花两性或稀单性，花冠黄绿色；果为蒴果、浆果或核果；种子多数有胚乳。主要分布在我国东北、西南、华中一带。柔嫩多汁，家畜喜食，宜饲喂马、牛、羊，不宜喂猪。

An annual or biennial herb, reproduced from seeds. It is commonly found in farmlands, and dam aging winter wheat, distributed in the southwestern, southern and central parts of the country. The feeding value is moderate, horses, sheep, goats and cattle like to graze fresh in spring.

276-2　猪殃殃　　276-1　猪殃殃花序

277 蓬子菜

学名 *Galium verum* L.
英名 Yellow Bedstraw

别名松叶草。多年生草本。地下茎横走，发达，暗紫红色。茎多直立，具四纵棱，多丛生。叶6~8片轮生，条形或狭条形，长1~3厘米，先端锐尖，边缘反卷，无柄，干枯后往往变为黑色。聚伞花序顶生和腋生，通常在茎顶集结成圆锥状，稍紧密。花小，黄色，花筒极短，4雄蕊伸出花冠以外。果片双生，近球形，无毛。全国各地均有分布，多生于空旷的草地或农区田埂，西北黄土高原多见。青鲜幼嫩时各种放牧家畜均喜食，干枯以后采食较少。

A perennial herb, commonly found in fieldsides, grasslands graveyards and wastelands, distributed in north, northeast, northwest China. Cattle, sheep and camel like to graze when green, the palatability is reduced if dry.

277-2 蓬子菜花序 **277-1 蓬子菜**

四十八 败酱科

278 异叶败酱

学名 *Patrinia heterophylla*
英名 Yellow patrinia

多年生草本。被短毛，茎很少分枝。基生叶有长柄，茎生叶略小，对生，羽状深裂，中央裂片最大。花顶生或腋生，呈聚伞花序，花黄色，花冠筒状，筒内有白毛。瘦果倒卵形。生于黄土高原东部或南部海拔较低处（1 000米）的山坡、路旁，适口性中等，幼嫩时放牧家畜喜食，一旦开花以后，适口性迅速下降。

278-2 异叶败酱花序

A perennial herb, spread by rhizome and by seeds. It is commonly found among hassocks along hill-slopes and roadsides, distributed in provinces of northeast China as well as Gansu, Shaanxi, etc. The feeding value is moderate. Cattle and sheep like to graze.

278-1 异叶败酱

279 缬 草

学名　*Valeriana officinalis*
英名　red valerian

别名败酱草。多年生草本。茎中空，有纵棱，被粗白毛。根状茎浓香。叶对生，多对羽状深裂，全缘或具极疏的锯齿，两面和柄部或多或少被毛。花为伞房状三出聚伞圆锥花序，苞片羽裂，花萼内卷，花冠紫红色或白色，筒状，瘦果。主要分布在甘肃、陕西，从东北至西南诸省均有分布，幼嫩时家畜采食，粗老后适口性下降。

A perennial herb, reproduced from seeds. It is commonly found on grasslands and forest, distributed in northeast and southwest China. The feeding value is low. Cattle and sheep graze it when young fresh.

279　缬草

四十九　桔 梗 科

280　桔 梗

学名　*Platycodon grandiflorus*
英名　balloon flower

多年生草本。高 40～90 厘米。体内含乳汁，全株光滑无毛。根粗大肉质，圆锥形或有分叉，外皮黄褐色。茎

280-2　桔梗花序

280-1　桔梗

直立，有分枝；叶多为互生，少数对生，近无柄，叶片长卵形，边缘有锯齿。花单生于茎顶或数朵成疏生的总状花序；花冠钟形，蓝紫色或蓝白色，裂片 5。蒴果卵形，熟时顶端开裂。喜生于松软肥沃的土壤，我国各地多有分布。牛、羊喜食。

A perennial herbaceous plant reproduced from seeds. It is commonly found in grassy hillsslopes and among shrubs in mountain areas, or grows in field-sides in low mountain areas. It is distributed in south and central Loess plateau as well as Sichuan and Qinghai. Sheep and cattle will graze it.

281 石沙参

学名 *Adenophora Polyantha* Nakai.
英名 Manyflower Ladybell

多年生草本。具白乳汁。茎直立。茎叶互生，无叶柄或具极短柄，叶片条形或条状披针形，边缘有尖锐的锯齿。花序较长，总状，有疏生分枝而呈圆锥状，生于茎顶。花冠蓝紫色，宽钟形，无毛，5浅裂。花萼裂片5，裂片尖，边缘有刺。蒴果倒卵形，种子卵形，黄棕色。羊极喜食，其他家畜食性一般，稍粗老则适口性下降，家畜微嗜。

A perennial herb, it grows grasslands. It is distributed throughout the Loess Plateau as well as many other provinces such as Hebei, Anhui, Jiangsu, etc. The palatability is poor, cattle and sheep graze its leaves and branches.

281 石沙参

282 柳叶沙参

学名 *Adenophora gmelinii* (Spreng.) Fisch.

多年生草本。主根肥大，入土深。茎直立，圆柱形，有纵棱，被疏毛；分枝下少上多，花后下部易变紫红色。叶片呈柳叶状，互生，无柄或具短柄，先端渐尖，基部楔形。总状花序，有分枝，较长，生于枝顶；花冠蓝紫色，宽钟形似呈喇叭状，光滑无毛。蒴果卵圆形，种子土黄色。生于林缘草丛、沟谷坡地，分布在我国东北、华北、内蒙古、甘肃、陕西等地。中嗜牧草，幼嫩时牛、羊喜食。

A perennial herb, it exists in grassy hill-slopes, under forest and among thickets in mountain area or field-edge. It is distributed in north, northwest and northeast China as well as southeastern Loess plateau. The feeding value is moderate and animals will graze.

282-2 柳叶沙参叶

282-1 柳叶沙参

283　秦岭沙参

学名　*Adenophora petiolata*
英名　Qinling Ladybell

多年生草本。根柱形,具稀疏柔毛,基生叶柄长,上部叶无柄。上叶片卵圆形,先端急尖,基部心形,边缘具锯齿,背面沿脉疏生短毛；茎生叶互生。花序不分枝呈总状,或下部有分枝成圆锥状,苞片披针形,先端渐尖,边缘具疏锯齿,小苞片 1～2,全缘；花萼无毛,倒圆锥形,裂片披针形或狭三角形,花冠蓝色或白色,裂片三角形。花柱与花冠近等长或稍伸出。果实倒卵圆形,果皮膜质。喜生于山坡草丛、崖边,分布在秦岭及毗邻的黄土高原沟壑区。牛、羊喜食其嫩枝叶,花后易粗老,适口性快速下降。

A perennial herb, it is commonly found in grassy hill-slopes, under forest and among thickets in Qinling mountain area. It is distributed in southeastern Loess plateau, mainly exists in Gansu and Shaanxi provinces. Cattle and sheep like to graze its leaves and fresh branches.

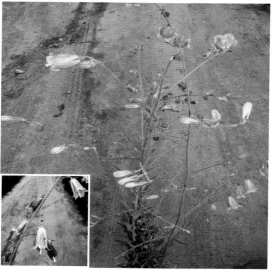

283-2　秦岭沙参花序　　283-1　秦岭沙参

五十　菊　科

284　串叶松香草

学名　*Silphnum perfoliatum* L.
英名　Perfoliate Rosinweed

多年生草本。根粗壮,有多节的水平根茎。茎秆粗壮,直立,四棱形,幼嫩时被毛。叶片长椭圆形,基生叶有柄,茎生叶无柄。叶面轻微的波状皱缩,叶缘有缺刻,叶面和叶缘疏生硬毛。伞房花序着生于假二叉分枝的顶端,花杂性,外缘 2～3 层属雌性花,花盘中央为两性花。瘦果心形,扁平,褐色,边缘具翅。1979 年我国从朝鲜引入,1984 年在庆阳黄土高原试验站种植。其基本特点是喜水肥,能高产,适口性差。由于茎秆木质化程度高,叶片具硬毛等缺点,宜幼嫩时打浆或切碎饲喂；调制成干草粉与其他牧草混合饲喂亦可。

An erect perennial herb reproduced from seeds. It is cultivated in lands of fertility and get high yield, but palatability is not good. Cattle and sheep will graze when it is fresh in early spring. Frequent cutting will improve quality and increase animal intake.

284-1　串叶松香草

284-2　串叶松香草花序

285 野茼蒿

学名　*Crassocephalum crepidioides*
英名　Hawksbeard Velvetplant

别名革命菜、山茼蒿、孔明菜。一年生草本。茎直立，少分枝，无毛或稀被柔毛。叶椭圆形或长椭圆形，顶端尖，基部楔形，边缘有不规则的锐齿。叶片基部羽状分裂，侧裂1~2对，叶柄柔软，具极狭的翅，茎上部叶小。头状花序排列成顶生或腋生的伞房花序。总花梗较长，总苞钟形，小花全管状，花冠橙红色。瘦果圆柱状，紫红色，具纵肋，冠毛白色。喜生于河边、路边、荒地，也入侵农田。幼嫩多汁，各种家畜均喜食。

An annual herbaceous plant, commonly found in farmlands, wastelands and ditch sides under forest. It is distributed in Gansu, Shaanxi and south of the Yangtze River. It is an excellent forage for the most livestock animals and can be used as vegetables.

285　野茼蒿

286 鬼针草

学名　*Bidens pilosa* L.
英名　Railway Beggarticks

一年生草本。茎直立，植株较小花鬼针草为高。中部叶对生，叶3深裂或羽状分裂，裂片卵形或卵状椭圆形，边缘有锯齿；上部叶对生或互生，3裂或不裂。头状花序多数单生于分枝顶端，总苞基部有细软毛。舌状花黄色或白色，筒状花黄色。瘦果长条形，有4棱，先端有3~4芒状冠毛。分布较广，农田及其周边多见，家畜喜食。

An annual herb reproduced from seeds. It is one of the noxious weeds in China, distributed throughout the country. The nutritive value is high, can be edible when it is young, and animals like to graze, but the seeds with thorn are bad for wool of sheep.

286　鬼针草花序

287 小花鬼针草

学名　*Beggarticks paruiflora* Willd.
英名　Smallflower Beggarticks

别名细叶刺针草、小刺叉、小鬼叉、一包针。一年生草本。小花鬼针草与鬼针草的主要区别是叶互生，下部叶柄长，具毛，多为 2 回

287-1　小花鬼针草

287-2　小花鬼针草花序

羽状 5 深裂，大裂片先端渐尖，基部下延成两侧带翅短柄，中裂片最大。青鲜时家畜喜食。开花结实后因果实针状，具倒刺毛，适口性下降且果实易黏结羊毛，甚至刺伤皮肤。

An annual herb reproduced from seeds. It is exists in water sides, wastelands and roadsides, distributed in provinces of north, and northeast China as well as Gansu, Shaanxi, etc. The feeding value is moderate. Cattle, sheep and goats are readily to graze when fresh.

288 腺梗豨莶

学名　*Siegesbeckia pubescens* Makino.
英名　Glandularstalk St.Paulswort

别名唐本草。一年生草本。茎直立，上部有分枝，通体被白色柔毛。叶对生，叶片宽卵形或卵状三角形至卵状披针形，基部楔形下延成翅状叶柄，两面有毛，边缘具齿。头状花序多数，排列成圆锥状。花序梗和总苞都有头状有柄的腺毛，分泌黏液。花黄色，边花舌状，心花筒状。瘦果楔形，黑色，上面有褐色瓣。喜沟谷湿润地。适口性差，放牧家畜不喜食，但调制成干草或青贮草，采食性增强。

An erect annual herb reproduced from seeds. It is commonly found in wastelands, roadsides, damp low lands in the mountain valleys and farmlands. It is distributed in south and southeastern Loess Plateau. Domestic animals do readily to graze.

288-1　腺梗豨莶

288-2　腺梗豨莶花序

289 飞廉

学名 *Carduus crispus* L.
英名 Curly Bristle Thoistle

越年生草本。茎直立,粗壮,有分枝,具条棱,有绿色翅,翅上有翅刺。一般第一年生长基生叶,叶片长大,有短柄,羽状深裂,主脉强大而明显,边缘波状具硬刺。茎生叶互生,无柄,从下向上逐渐变小。头状花序1~4个簇生于枝顶。总苞钟状,多层,先端成刺,向外反曲。花筒状,紫红色。瘦果长圆柱形,稍弯曲,冠毛白色或灰白色,刺

289-4 飞镰幼苗　　**289-2** 飞廉花序　　**289-3** 飞廉茎　　**289-1** 飞廉

毛状。西北常见植物,主要分布在沟坎低洼处,适口性差,家畜可食。

A biennial herb reproduced from seeds.It is common in wet-lands, roadsides and farmlands.It is a very bad weed of old orchards and alfalfa.It is distributed throughout the country.The feeding value is low.Cattle, sheep, goats and horses grazing are limited when mature.

290 天名精

学名 *Carpesium abrotanoides* L.
英名 Common Carpesium Fruit

别名天蔓青。多年生草本。茎直立,中部以上多分枝,枝条密生短柔毛。叶互生,上部叶片较小,无柄;下部叶片大,宽椭圆形或长圆形,先端尖或钝,基部有具翅的叶柄;叶片全缘或先端具齿,叶背有短毛和腺点。头状花序多数,贴茎枝腋生,具短梗或无梗。总苞钟状球形,总苞片3层,外层短,无毛。花黄色,外围的雌花花冠丝状,中央的两性花花冠筒状。瘦果条形,有细纵棱,褐色,先端有喙,具腺体。开春幼嫩时牛、羊乐食,开花后不食,冬季可食。

An erect perennial herb reproduced from seeds.It is commonly found in hill-slopes, lawns, farmlands and roadsides, distributed throughout the country.The feeding value is moderate.Sheep like to graze it fresh, but cattle do not.After it is withered sheep, goats and cattle like to graze it.

290-1 天名精　　**290-2** 天名精花序

291 刺儿菜

学名 *Cephalanoplos Segetum* (Bege.) Kitam
英名 Little Thistle

别名小蓟。多年生草本,根状茎长。茎直立,有毛或无毛。叶椭圆或椭圆状披针形,先端尖,基部圆,全缘或有齿裂,具刺,两面密生蜘蛛丝状的绵毛。雌雄同株或异株,头状花序单生于茎顶,总苞多层,具刺。管状花,紫红色。瘦果,冠毛羽状。切碎或打浆后可喂猪,幼嫩时马、牛、羊采食少,干制后稍作加工即为家畜的优等饲料。

It is a perennial herb that propagates mainly by crown buds. It is commonly found in farmlands, damages wheat and other dry-land crops. It is distributed throughout the Loess Plateau. Cattle, sheep and goats like to graze after an autumn frost.

291-2 小蓟花序   291-1 小蓟

292 大 蓟

学名 *Cephalanoplos Segetum* (willd.) kitam
英名 Setose Cephalanoplos

别名大刺儿菜。多年生草本,有长匍匐根。茎直立,上部有分枝,全株被蛛丝状毛。叶互生,具短柄或无柄。叶片长圆形,边缘有缺刻状齿或浅裂,有细刺,叶背蛛丝状毛多于叶面。头状花序较小,数朵集生于茎的上部。花单性,雄花序较小,雌花序略大。总苞片多层,外层短。花冠淡红紫色,全为筒状花。瘦果长圆形,黄白色或淡棕色,冠毛羽状。适口性较差,但秋季经霜后,家畜喜食。

292-3 大蓟幼苗   292-2 大蓟花序   292-1 大蓟

A perennial herb, spread by crown buds and by seeds. It is commonly found in wastelands and farmlands, damages winter wheat, soybeans and some other crops. It is distributed throughout the Loess Plateau and over China. Cattle, sheep and goats, even pigs, are readily to graze in the spring and autumn.

293 菊苣

学名 *Cichorium intybus* L.
英名 Common Chicory

多年生草本。基生叶莲座丛状,倒披针状椭圆形,基部渐狭有翼柄,羽状深裂或不分裂,侧裂片镰刀形。花茎直立,具条棱,分枝扁斜且顶端粗厚,疏被糙毛或无毛;茎生叶渐小而少,披针状,叶两面疏被长节毛。头状花序单生于枝端或数个簇生于上部叶腋,总苞圆柱状,总苞片2层,披针状。花冠蓝色,舌状。瘦果顶端截形,冠毛短,膜片状。喜水肥,生长快,富含营养,适口性好,各种畜、禽皆喜食。

A perennial herb, spread by crown buds and by seeds. It grows in farmlands, creek sides and grasslands. It is distributed in north, northeast and northwest China. The feeding value is high. sheep, goats, rabbits and pigs like to graze it.

293-2 菊苣花序 293-1 菊苣

294 小飞蓬

学名 *Conyza canadensis*
英名 Canadian Fleabane

别名小白酒草。一年生或越年生草本。茎直立,坚硬,高大,有细条纹及粗糙毛。叶互生,具不明显的叶柄。叶片条状披针形或长圆状条形,全缘或具微锯齿,有长睫毛。头状花序具短梗,多数密集成圆锥状或伞房形圆锥状。总苞半球形,苞片2~3层,条状披针形,边缘膜质。舌状花小而直立,白色,筒状花比舌状花稍短。瘦果长圆形,冠毛刚毛状,污白色。主要生长于河滩、渠旁及灌溉农田,有时在稀疏的庄稼地和路边积过水的地方可见大片群落。幼嫩枝叶家畜喜食,若适时调制成青干草,是家畜越冬的好饲草。

An annual or biennial herb reproduced from seeds. It occurs in river beaches, creek-sides, wastelands and farmlands. It is distributed in south and southeast Loess Plateau. The feeding value is moderate. Livestock like to graze it when fresh but very limited after flowering.

294 小飞蓬

295 旱莲草

295-1 旱莲草

295-2 旱莲草花序

学名 *Eclipta prostrata* L.
英名 Yerbadetajo

别名墨草、鳢肠。一年生草本。茎自基部和上部分枝,直立或匍匐,着地后节上易生根,绿色或红褐色,被伏毛。叶对生,无柄或基部有短柄,被粗伏毛。叶片长披针形、椭圆状披针形或条状披针形,全缘或具细齿。头状花序有梗,顶生或腋生。总苞片6枚,有毛,宿存,边花舌状,雌性,白色;心花筒状,两性。舌状花的瘦果四棱形,筒状花的瘦果三棱形,表面有瘤状突起,无冠毛。生于水沟边或潮湿地。中等牧草,牛、羊乐于采食。

An annual legume, reproduces from seeds. It is commonly found in wet lands and creek. It is distributed over Loess Plateau and throughout the country. It is palatable for cattle and sheep in autumn, winter and spring.

296 一年蓬

学名 *Erigeron annuus* (L.) Pers.
英名 Annual Fleabane

别名千层塔。一年生或越年生草本。全株被短毛。茎直立,上部有分枝。基生叶长圆形或宽卵形,边缘有粗锯齿,基部渐狭变成带翅的叶柄;中上部叶片较小,披针形或长圆状披针形,边缘有不规则的锯齿,有柄或无柄;最上部叶片条形,全缘。头状花序排列成伞房状或圆锥状,边花舌状,白色或淡蓝色;心花筒状,黄色。瘦果长圆形,扁,有一层很短的鳞片状冠毛。该草生于较湿润的草地或农田周边,家畜喜食,尤其是经秋霜后,适口性大增。

An annual or biennial herb reproduced from seeds. It is commonly found in wastelands, roadsides and farmlands. It is a serious weed of cotton, pulses and small grains. It is distributed in southeastern Loess Plateau as well as other provinces such as Henan, Sichuan, etc. The feeding value is moderate and can be used by ruminants in autumn.

296-2 一年蓬花序   296-1 一年蓬

297 辣子草

学名　*Galinsoga parviflora*
英名　Herb of Smallflower Galinsoga

别名向阳花。一年生草本。茎直立,多分枝,全株被柔毛。叶对生,具柄。叶片卵圆形至披针形,先端渐尖,基部圆形至宽楔形,边缘具毛,有微齿,叶基三出脉明显。头状花序较小,有细长梗。总苞半球形。舌状花白色,筒状花黄色,花托有披针形托片。瘦果圆锥形,有棱和向上的刺毛,冠毛鳞片状。主要生长在果园、空地、路旁及水肥较好的农田周边。适口性一般,家畜可食。

An annual herb reproduced from seeds. It occurs in farmlands and barren slopes, distributed in southwest China and Shaanxi province. The feeding value is moderate. Cattle, buffalo, sheep and goats like to graze in autumn and winter.

297-2　辣子草花序　　297-1　辣子草

298 泥胡菜

学名　*Hemistepta lyrata* Bunge
英名　Lyrate Hemistepta

二年生或一年生草本。直根系,主根发达,侧根少,入土不深。茎直立,疏被白色蛛状绵毛。基生叶莲座状,通常在花期早枯;中下部叶片倒披针形或椭圆状披针形,大头羽状深裂或近全裂,顶裂片较大,三角形,侧裂片7~8对,长椭圆状倒披针形,叶面绿色,叶被具灰白色绒毛;上部叶片条状或条状披针形。头状花序多数,在株上部排列成疏伞房状,总苞球形,总苞片多层,背面顶端有紫红色鸡冠状附片。花紫红色,全为管状花。瘦果圆柱形,冠毛白色,两层,羽状。多生于山坡草地、路旁,有时入侵农田和果园。优良牧草,茎叶柔嫩,家畜喜食。

A biennial herb reproduced from seeds. It occurs in wastelands and farmlands. It is distributed throughout the country except Xinjiang and Tibet. The feeding value is high. Cattle, sheep, goats and pigs will graze before flowering in spring.

298　泥胡菜

299 蓼子朴

学名　*Inula salsoloides* Ostenf.
英名　Salsola-like Inula

多年生草本,靠根芽和种子繁殖。根横走,茎丛生,多分枝,直立或斜生披散。叶互生,无柄;叶片披针状或长圆状条形,半肉质,先端尖,基部宽,半抱茎,全缘,叶面无毛,叶背有短毛和腺点。头状花序单生于枝顶,总苞片3~4层,内长外短;边缘舌状花先端3裂,中间筒状花先端5裂。瘦果近长圆形,有毛,有细棱,褐色。广布于"三北"地区,为荒漠、沙地多见。在沙漠周边容易形成大面积群落覆盖,当地放牧家畜喜食。

299-1　蓼子朴

299-2　蓼子朴花序

A perennial herb, spread by crown buds and by seeds. It occurs in sandy lands and grasslands, distributed in north and northwest China. The feeding value is low. Camels like to graze fresh or dry. Sometimes sheep would graze it in winter.

300 山苦荬

学名　*Ixeris chinensis*
英名　Chinese Lettuce

多年生草本。全株含乳汁,无毛,有匍匐根。茎直立或半直立,基部或叶腋分枝。基生叶丛生,条状披针形,先端渐尖,基部变狭成窄叶柄,有不规则的羽裂或锯齿;茎生叶互生,下部大上部小,细而尖,无柄,半抱茎。头状花序排列成稀疏的聚伞状,总苞在花未开时呈圆筒状。花全为舌状花,多数为黄色,也有白色的。瘦果长椭圆形或纺锤形,稍扁,有条棱,棕褐色,具长喙,冠毛白色。品质优良,家畜喜食。

A perennial herb, spread by crown buds and by seeds, usually found in dry-land crops. It is distributed throughout the Loess Plateau. The feeding value is quite high. Sheep, goats, pigs, rabbits and poultry like it.

300　山苦荬

301 抱茎苦荬菜

学名 *Ixeris sonchifolia* Hance.
英名 Sowthiotle-leaf Lxeris

多年生草本。通体含乳汁，无毛。茎直立，顶部分花枝。基部叶丛生，长圆形，边缘具不整齐的羽裂，羽裂处全缘或具齿；茎生叶较小，互生，卵状长圆形或舌状卵形，先端尖，基部宽大并变成耳状或戟状抱茎，全缘或羽状分裂。头状花序密集成伞房状，花全为舌状花，黄色。瘦果纺锤形，有喙，冠毛白色。生长于山坡、沟谷、田埂、路边及空闲地。优质牧草，各种家畜均喜食。

A perennial herb that occurs in hill slopes, riversides and sparse weeds. It is distributed in provinces of north and northeast China as well as Gansu and Shaanxi provinces, etc. Palatability is good and feeding value is high. Various domestic animals and poultry like it.

301-2 抱茎苦荬菜叶　　301-1 抱茎苦荬菜

302 蒙山莴苣

学名 *Lactuca tatarica*.
英名 Tartar Lettuce

别名紫花山莴苣。多年生草本，通体含乳汁，无毛。具直根和匍匐根，茎直立，有分枝或无分枝。基生叶基部半抱茎，叶片长圆形，羽状或倒向羽状深裂或浅裂，叶缘和裂齿先端刺状；中部叶裂片渐少，上部叶全缘或仅具小齿，无柄。头状花序多数在茎枝顶排列成疏圆锥状。总苞圆筒状，苞片里长外短，紫色或绿色。花全为舌状花，蓝紫色。瘦果长椭圆状条形，具纵棱，果颈较长，冠毛白色。优良牧草，家畜、家禽都喜食。

A perennial herb, spread by crown buds and by seeds. It is commonly found in riversides, lakesides and farmlands, distributed in northern Yellow River. The feeding value is high. Cattle, sheep, goats, rabbits and pigs like to graze it.

302-1 蒙山莴苣　　302-2 蒙山莴苣花序

303　兴安毛莲菜

303-1　兴安毛莲菜

303-2　兴安毛莲菜苗

303-3　兴安毛莲菜花序

学名　*Picris davurlca.*
英名　Oxtongue

多年生草本。直立，通体有刺，叶互生，披针形，具短柄或无柄。边缘有针状刺。头状花序多数在茎枝顶排列成疏圆锥状。总苞圆锥状，苞片长三角形。舌状花与管状花黄色。在我国广有分布，主要生长于林下灌丛、林间荒地及农田周边。适口性良好，无论青草、干草家畜均喜食。

A perennial erect herb, reproduced from seeds, It is commonly found in shady side of the wastelands, orchards. It is distributed throughout the country except southern China. The feeding value is low with sheep and goats like to graze dry leaves in winter.

304　阿尔泰狗娃花

学名　*Heteropappus altaicus* Novopokr.
英名　Altai Heteropappus

别名阿尔泰紫菀。多年生草本。全身具弯曲短硬毛和腺点。多数茎由基部分枝斜升。叶疏生或密生，条形、披针形或匙形；先端尖或钝，基部楔形，全缘。头状花序单生于枝顶端或排列成伞房状，总苞片草质，舌状花淡蓝紫色，管状花黄色。瘦果长圆状倒卵形，冠毛粗，污白色或红褐色。该草在干旱草原、荒漠草原、草甸草原及农区到处可见。幼嫩时家畜喜食，随着生长粗老，适口性下降，干燥后则乐于采食。

A perennial herb reproduced from seeds. It is commonly found in grasslands, sand-lands and newly reclaimed farmlands, distributed in provinces of north and northeast China as well as Gansu, Shaanxi, etc. The feeding value is high. Cattle, sheep and horses are readily to graze in autumn and winter.

304-2　阿尔泰狗娃花花序　　304-1　阿尔泰狗娃花

305 裂叶马兰

学名 *Kalimeris incisa* (Fisch.) DC.
英名 Indian Kalimeris

别名鸡儿肠、田边菊。多年生草本。具根茎，茎直立，有分枝。叶互生，无柄。叶片倒披针形或倒卵状长圆形，先端钝或尖，基部渐狭。边缘多羽状浅裂或粗齿，上部叶渐小，全缘。头状花序单生于枝顶，排列为疏伞房状；总苞2~3层，边缘膜质，有睫毛；边花1层，舌状，淡蓝紫色；心花筒状，黄色。瘦果楔状长圆形，极扁。冠毛不等长，易脱落。种子和根茎均可繁殖。分布广泛。春秋两季家畜喜食，有的地方用来喂猪。

It is a perennial herb, spread by seed and by rhizome. It is commonly found in roadsides, farmlands and creek-sides, distributed throughout the country. It is excellent forage. It is palatable for livestock before flowering, Cattle and buffalo like to graze it in autumn.

305-2 裂叶马兰花序　　305-1 裂叶马兰

306 花花柴

学名 *Karelinia caspica*
英名 Caspian Sea Karelinia

别名胖姑娘。多年生草本。茎直立，多分枝。叶互生，无柄；叶片长圆状卵形或长圆形，先端圆钝或急尖，基部有圆形或戟形小耳，抱茎，全缘或有不规则的浅裂齿，被糙毛或无毛。头状花序数个疏生于枝顶。总苞片多层，内长外短；花托平，有托毛；花红紫色或黄色，边花雌性，花冠丝状，冠毛1层，心花两性，花冠细筒状，冠毛多层。瘦果圆柱形，冠毛白色，种子和根芽繁殖。生于沙漠和戈壁边缘，骆驼、山羊和绵羊均可采食。

A perennial herb, it is distributed in desert steppes of the north and northwest Loess Plateau. The feeding value is low, the palatability is poor, camels like to graze it all year around but horse do not like it any time of year, the other animals like to graze when it dry.

306-2 花花柴花序　　306-1 花花柴

307 风毛菊

学名 *Saussurea japonica*
英名 Japanese Saussurea

多年生草本。茎直立，上部分枝，有短微毛和腺点。基部叶和茎下部的叶有柄，叶片长圆形或椭圆形，羽状分裂，中裂片长圆状披针形，侧裂片狭长圆形，两面有短微毛和腺点。茎上部叶渐小，椭圆状、披针形或条状披针形，羽状分裂或不裂。头状花序在茎枝顶部排列成密伞房状；总苞筒形，被蛛丝状毛，总苞先端附属物往往带有紫红色。花冠全为筒状，紫红色。瘦果近楔形，褐色，冠毛淡褐色。多分布在山坡、灌丛或田埂上。幼嫩时家畜喜食，开花后粗老，适口性变差。

A perennial herb reproduced from seeds. It is common found in grasslands or among shrubs along hill slopes. It is distributed in north, northeast, northwest, east and south China. Horses, cattle and sheep are readily graze it in spring and autumn.

307-2 风毛菊花序

307-1 风毛菊

308 鳍蓟菊

学名 *Olgaea leucophylla*
英名 Whiteleaf Olgaea

别名白山菊、白背、火煤草。多年生草本。茎直立，粗壮，密被灰白色蛛丝状绵毛，不分枝或上部少分枝。叶长椭圆形或椭圆状披针形，先端具针刺，基部沿茎下延成翅，边沿具不规则的疏齿或为羽状浅裂，裂片、翅端及叶缘均具不等长的针刺，两面几乎同是灰白色，被蜘蛛丝状绒毛。头状花序单生于枝端或有时在枝端侧生1~2个小花序；总苞钟状，总苞片多层，先端具长刺；管状花粉红色，瘦果矩圆形，苍白色，具隆起的纵纹和褐斑。主要分布在黄土高原，喜沙耐旱，散生于多种草原。幼嫩时绵羊、山羊采食，因具刺太多，适口性较差，饲用价值不高。

A perennial herb reproduced from seeds. It grows on grasslands, wastelands and roadsides, distributed all over the Loess Plateau. The feeding value is low with goats and sheep grazing when it is tender. Camels and cattle like to graze its dry inflorescence in autumn and winter.

308-1 鳍蓟菊   308-2 鳍蓟菊花序

309 顶羽菊

学名 *Acroptilon repens* L.
英名 Creeping Acroptilon

别名苦蒿、牛毛蒿、灰叫驴。多年生草本。根粗壮,茎直立,高 25~70 厘米,多分枝,有纵沟棱,被淡褐色毛。叶无柄,叶片条形、披针形或长椭圆形,先端尖,全缘或有锯齿,两面具灰色绒毛或蛛丝状毛和腺点。头状花序单生于枝端,总苞长椭圆形或头状乱形,总苞片多层,呈覆瓦状排列,上半部透明膜质,花冠红紫。瘦果倒长卵形,冠毛白色。耐旱,多生于荒坡草地,也入侵农田和果园。该草具苦味,适口性较差,青绿时牛采食,羊采食很少。经霜以后,多种家畜均采食。

309-2 顶羽菊花序　　309-1 顶羽菊

A perennial herb, reproduced from seeds. It occurs in arid regions of mountain area, wastelands and fields sides. It is distributed in Gansu, Shaanxi and north, northwest China. Goats like to graze tender branches and other animals do nott graze when fresh.

310 野　菊

学名 *Dendranthema indicum* L.
英名 Mother Chrysanthemum

有的地方叫"九月菊"。多年生草本。茎多为斜升,基部和上部均多分枝。茎生叶卵形或矩圆状卵形,羽状深裂,边缘浅裂或具齿;叶面有腺体,疏生柔毛,叶柄具翅。头状花序排列成伞房状圆锥花序,小花数量多,雌性的舌状花和两性的筒状花均为黄色,花期长,花色鲜艳。瘦果无冠毛。生于干旱山坡、地埂、崖边,幼嫩时家畜喜食,经霜后特别喜食。

A perennial herb. It occurs in hill-slopes, grassy slopes and wastelands. It is distributed in the gully of the Loess Plateau as well as eastern China. The feeding value is low and sheep and goats do not like it, but would graze after a frost or dry in autumn and winter.

310-1 野菊　　310-2 野菊花序

311　蒲公英

学名　*Taraxacum officnala*
英名　Mongolian Dandelion

多年生草本。全株含白色乳汁。根圆锥形，叶莲座状平展；倒披针形，羽状深裂，顶裂片较大，全缘或具波状齿。花葶数个，上部密被蜘蛛丝状毛；头状花序单生，外层总苞片短，卵状披针形，内层条状披针形，顶端有小角；舌状花黄色。瘦果倒披针形。分布广泛。着生于天然草地，农田、路旁、渠边及荒地。常用作放牧鲜食，蛋白质含量高，营养丰富，适口性好，多种家畜喜食，尤其是猪禽类。

A perennial herb, commonly found in roadsides, farmlands and orchards. It is distributed in various regions in the country except south China. The feeding value is quite high. It is palatable for all animals, including poultry. In addition, the leaves are edible when it is fresh.

311-1　蒲公英

311-2　蒲公英花序

312　苦荬菜

学名　*Ixeris dentate* Nakai (*Sonchus Oleraceus* L.)
英名　Common Sowthistle

别名苦菜。多年生草本。茎直立，无毛。基生叶倒披针形，先端尖锐，基部下延成叶柄，边缘有疏锯齿或呈羽状分裂；茎生叶披针形，基部略呈耳状，无叶柄。头状花序多数，有细梗，排成伞房状。外层总苞片小，卵形，内层总苞片条状披针形；舌状花黄色。瘦果纺锤形，黑褐色，冠毛浅棕色。主要分布在农耕区，喜温暖水肥条件好的地方。适口性良好，各种家畜和多种家禽都喜食。

A perennial herb, spread by crown buds and by seeds. It is common found in farmlands and wastelands, damages various kinds of farm crops. It is distributed over all parts of the country. The nutritive and feeding value is quite high. It is edible and various animals and poultry like it.

312-1　苦荬菜

312-2　苦荬菜花序

312-3　苦荬菜叶

· 156 ·

313 苍 耳

学名　*Xanthium sibiricum* Patrin.
英名　Siberian Cocklebur

一年生草本。茎直立，粗壮结实，多分枝，其上有钝棱及长条状斑点。叶互生，具长柄。叶片三角状卵形或心形，基出3脉明显，边缘浅裂或具齿，叶两面均有糙毛。花单性，雌雄同株，雄头状花序球形，密集于枝顶；雌头状花序椭圆形，生于雄花序下方。总苞有钩刺，内含2花，成熟时总苞变硬，内含2瘦果，倒卵形。家畜可食，制干后喜食。因果实上有钩刺，容易黏结羊毛，所以在种子成熟前后不宜作放牧。

313-1　苍耳　　　313-2　苍耳幼苗

An annual herb reproduced from seeds. It is commonly found in wastelands, roadsides and creek banks. It is distributed over all parts of the country. Most domestic animals like to graze its tender branches and leaves. It can be easily stick to the wool as its seed surface has thorns.

314 款 冬

学名　*Tussilago farfara*
英名　Common Coltsfoot Flower

菊科多年生草本。具有十分发达的横走根茎，新根茎白色，脆嫩，节上生根长芽；老根茎土黄色，较柔韧，多数花蕾在秋季形成于地下，着生于老根茎。冬季花茎先叶出土，开花甚早。头状花序顶生，周围舌状花，中间管状花，均为黄色。叶丛生柄长；叶片心形或阔卵性，边缘有波状疏锯齿，叶背及叶柄密生丝状白色长绵毛。瘦果黄白色，具白色长冠毛，种子和根茎均可繁殖。主要分布在黄土高原的泾河、渭河、北洛河流域流水及潮湿的沟底部。鲜嫩时家畜少食或不食，叶枯萎后牛、羊采食。

A perennial stem-less herb, propagated mainly by rhizome and by seeds. It is commonly found in wet place of gullies of the Loess Plateau. When it is fresh, most animals do not like to graze but the sheep and goats will consume dry leaves in autumn, winter and the following spring.

314-3　款冬花蕾　　　314-1　款冬　　　314-2　款冬花序

315　黄鹌菜

315-1　黄鹌菜

315-2　黄鹌菜花序

学名　*Youngia japonica* L.
英名　Japanese Youngia

一年生草本。茎直立,不分枝,株色常呈暗紫色,茎光滑无毛。基部叶丛生,倒披针形,羽状深裂或浅裂,顶裂片比侧裂片大。幼苗叶片边缘有不规则的疏齿,叶柄具翅。茎生叶互生,通常两三片,小。头状花序在茎枝顶部排列成聚伞状圆锥花序。总苞钟状,外苞片5,三角形或卵形,内苞片8,披针形,舌状花黄色。瘦果长圆状椭圆形,有粗细不等的纵棱,冠毛白色。优质牧草,适口性好,家畜喜食。

An annual herb reproduced from seeds. It occurs in wilderness or farmlands, distributed throughout the Loess Plateau except north and northwest. The feeding value is high. It is palatable for all animals, including poultry.

316　大丁草

学名　*Gerbera anandria* L.
英名　Gerbera

多年生草本。植株有春秋二型。春季型高5~10厘米,秋季型高30厘米。春季型叶小,秋季型叶大。叶片提琴状大头羽状分裂,背面及叶柄密生白色绵毛,后随生长逐渐脱落。头状花序单生,春季型有舌状花和筒状花两种,秋季型仅有筒状花。瘦果,冠毛污白色,春季型5~6月开花,秋季型8~9月开花。生于林缘、林下、路旁、田边等地。家畜喜食。

316-1　大丁草

316-2　大丁草花序

A perennial herb reproduced from seeds. It occurs on mountain area, gullies, among thickets and under forest, distributed throughout the country. Animals like to graze the whole plant, the pigs like it after boiled.

317　火绒草

学名　*Leontopodium smithianum*
英名　Common Edelweiss

多年生草本。茎数条簇生，分可育茎和不育茎，通体被白色柔毛。叶互生，无柄，先端渐尖，基部渐狭。叶背密被白色柔毛较叶面多。头状花序密集成无柄的伞房状，总苞密被白色长绵毛，苞叶数片条状披针形。瘦果，具短粗毛，冠毛白色。该草耐旱性强，多生于山坡草地或农田地埂，草质较差，只有在幼嫩和经霜后家畜采食。

A perennial herb reproduced from seeds.It grows in typical steppes, mountain meadows and sandy grasslands and distributed over Loess Plateau.Animals would graze it fresh but prefer it dry in the winter.

317-2　火绒草花序　　　317-1　火绒草

318　秋鼠麴草

学名　*Gnaphalium hypoleucum* D.Don
英名　Autumn Cudweed

一年生草本，种子繁殖。茎直立，常簇生，不分枝或少分枝，密生白色绵毛。叶互生，无柄或叶片下延成柄，耳状抱茎；叶片条形或条状披针形，先端有小尖，基部渐狭成翅，全缘，两面有灰白色绵毛。头状花序多数，通常在顶端密集成伞房状；总苞球状钟形，黄色，干膜质；

318-1　秋鼠麴草　　　　318-2　秋鼠麴草花序

花黄色。瘦果矩圆形，冠毛污白色。生于荒地、草坡、沟谷等较湿润的地方。一般情况下牛、羊不采食，秋季霜后枯萎或冬季缺草时家畜采食较好。

An annual herbaceous plant reproduced from seeds.It occurs in grasslands of hill-slopes, forest fringes and farmlands.It is distributed in east, central, south and southwest China as well as Gansu, Shaanxi and Henan.The feeding value is low, livestock will graze it when not enough feed.

319 伞花雅葱

学名　*Scorzonera albicaulis* Bunge.
英名　whitestem Serpentroot

319-1 伞花雅葱

319-2 伞花雅葱花序

别名笔管草。多年生草本。全株密被蛛丝状绵毛，后渐无毛或无毛，仅上部有毛。茎直立，具沟槽，基部被残叶鞘，上部分枝。基生叶线形，有时超出茎，基部渐狭成柄，柄基部扩大成鞘，淡褐色，先端渐尖，具5~7条平行脉，表面无毛，背面疏被柔毛；茎生叶与基生叶同形，较小，无柄、抱茎，向上渐小。头状花序数个，排成伞房状；总苞狭筒形，初被蛛丝状绵毛，后渐脱落，总苞片5层，覆瓦状排列，外层小，三角状卵形，先端锐尖，中层卵形或卵状披针形，内层线状披针形，具宽膜质边；舌状花黄色，背面稍带淡紫色，超出总苞，先端5齿裂。瘦果无毛，黄褐色，圆柱形，具纵肋，先端渐狭成喙，稍弯；冠毛黄褐色，生于干山坡、固定沙丘、沙质地、山坡灌丛、林缘、路旁等处。分布于我国东北及黄河流域以北各省区。

A perennial herbaceous plant, commonly found on mountain grasslands, sandy lands and among hassocks or forest. It is distributed in northeast China and around the Yellow River. The feeding value is high and various livestock animals will graze it.

320 拐轴雅葱

学名　*Scorzonera divaricata* Turcz. var. *divaricata*
英名　Divaricate Serpentroot

多年生草本。根粗壮，圆柱形，暗褐色。根颈密被纤维状叶鞘残遗。茎通常单一或上部少分枝。基生叶条形、条状披针形或条状矩圆形，先端长渐尖，基部窄成翅状柄，边缘波状或稍呈波状皱曲；头状花序单生于枝顶，舌状花黄色，瘦果圆柱形，冠毛污白色。多散生于草原、山地草坡、林缘或草甸，分布在我国华北和西北。中等饲用植物，各种家畜均喜食，春季采食后能促进家畜体力恢复。

A perennial herbaceous plant, spread by crown buds and by seeds. It is commonly found in typical steppes and hill-slopes, and is distributed in north and northwest China as well as Loess plateau. The feeding value is moderate, cattle and sheep will graze it.

320-2 拐轴雅葱花序　　320-1 拐轴雅葱

321 旋覆花

学名　*Inula japonica* Thunb.
英名　Inula Flower

别名金佛草。多年生草本。茎直立,被长柔毛。叶互生,无柄;叶片长椭圆形至披针形,基部渐狭,半抱茎,全缘,叶背有疏伏毛和腺点。头状花序直径 2.5~4 厘米,通常数个排列成疏伞房状,稀单生,有细梗;总苞片多层,条状披针形,最外面一层较长,披针形,有毛;花黄色,边花舌状,先端 3~4 裂,心花筒状,先端 5 裂。瘦果圆柱形,褐色,有 10 条纵棱;冠毛白色,与筒状花等长。生长在水边、渠畔和路旁潮湿处。低等牧草,青嫩时家畜采食,粗老不食,调制干草后适口性提高。

A perennial herb reproduces from seed. It exists in water sides, ditch banks, damp roadsides and damp low farmlands. It is distributed in the northern, northeastern, central and eastern parts of China as well as Sichuan and Guangdong provinces, etc. Most domestic animals will like to graze it as hay.

321-2　旋覆花花序　　321-1　旋覆花

322 麻花头

学名　*Serratula centauroldes* L.
英名　Common Sawwort

多年生草本。茎直立,有棱。基生叶有柄,茎生叶通常无柄,基生的长叶柄往往残存在茎基部;叶片椭圆形,羽状深裂,裂片全缘或具疏齿;茎生叶裂片小,细条形。头状花序生于枝端,具长梗;总苞卵形,总苞片 5 层,顶端有膜质附属物,花冠淡紫色。瘦果,冠毛刚毛状,不等长,淡黄色。抗逆性强,多生于干旱山坡、地埂及荒草之中。适口性较差,幼嫩时采食,调制成干草家畜喜食。

A perennial herb that exists along roadsides, in wastelands and typical steppes. It is distributed throughout the Loess Plateau. The feeding value is moderate and livestock like to graze its tender leaves in the early stage of growth. Cattle, sheep, goats and horses like to graze its inflorescence in summer and autumn. Most domestic animals will like to eat it as hay.

322-1　麻花头　　322-2　麻花头花蕾

323 牛 蒡

学名 *Arctium lappa* L.
英名 Great Burdock

别名大力士、恶实。二年生草本。根肉质，纺锤形。茎直立，粗壮，高可达1米以上，具纵条纹，带紫色，有乳突状短毛，上部分枝多。基叶丛生，茎叶互生，宽卵形或浅心形，叶片大，先端钝圆，基部心形，全缘、波状或有细锯齿，叶背密被白色绒毛，有柄。头状花序多数或少数排列成伞房状，总苞卵球形，总苞片顶端呈刺钩状。花全部为筒状花，淡红紫色。瘦果，椭圆形，灰黑色。喜生于村边、河滩上、灌丛林下，特别是在过牧草地上会成片生长。幼嫩时猪和家兔喜食，其他家畜很少采食。子实和肉质根中含有丰富的蛋白质和脂肪，是很好的精饲料，各种家畜均喜食。

A biennial herb reproduced from seeds.It occurs in hill slopes, gullies, roadsides and nearby village, distributed throughout the county.The feeding value is low.It is poor for sheep as its fruit has hooks that can stick on wool.

323-1 牛蒡

323-2 牛蒡苗

324 菊 芋

学名 *Helianthus tuberosus* L.
英名 Jerusalem Artichoke

别名洋姜、鬼子姜。多年生草本。高2米左右，具块茎。茎直立，上部多分枝，密被短柔毛和刚毛。基叶对生，上部茎叶互生，叶片卵形、长圆状卵形或卵状椭圆形，先端尖或渐尖，基部楔形，边缘有锯齿，上面粗糙，下面有毛，叶柄上部具狭翅。头状花序多数，生于枝端，总苞片披针形，开展。舌状花淡黄色，管状花黄色。瘦果楔形，有毛，上端常有2~4枚具毛的短芒。菊芋适应性广，尤耐瘠薄，多有栽培种，块茎可作蔬菜，也可作多汁饲料。地上茎叶属优质饲草，各种家畜均喜食，特别是块茎喂猪效果良好。

An perennial erect herb, it is propagated by tubers, commonly cultivated in farmlands or wastelands. It is distributed throughout the Loess Plateau except north part. Sheep and cattle like to graze fresh and dry leaves, and the tubers can make quality livestock feed.

324-2 菊芋块茎   324-1 菊芋花序

325 漏芦

学名　*Rhaponticum uniflorum* (L.) DC.
英名　Radix Rhapontici

多年生草本。高 30～100 厘米；根状茎粗，根直伸；茎直立，不分枝，单一或丛生，被绵毛。基生叶和茎下部叶轮廓为长椭圆形或倒披针形，羽状深裂或近全裂，裂片 5～12 对，边缘有锯齿或呈 2 回羽裂状，两面被白色蛛丝状毛及黄色腺点；叶柄长 6～20 厘米；中上部的叶渐小，与基生叶同形，但无柄或仅具短柄。头状花序单生于茎顶；总苞半球形，总苞片约 9 层，向内层渐长，顶端各有宽卵形、褐色附属物；花夏秋开，紫红色，同型，全为两性管状花，花冠裂片长约为花冠管之半；花药基部箭形；花柱中部有毛环。瘦果楔形，具 3～4 棱。冠毛褐色，多层，不等长，基部联合成环。生于山坡丘陵地，常见于松林或桦木林下。分布于我国东北、西北、河北、河南、四川等省区。

325-2　漏芦花序

325-1　漏芦

A perennial herbaceous plant reproduced from seeds. It is commonly found in hill-slopes and among shrubs. It is distributed in north, northeast and northwest China. The feeding value is low, animals will graze fresh leaves in early spring and cattle can consume its inflorescences.

326 丝裂亚菊

学名　*Ajania nematoloba*
英名　Threadylobe Ajania

多年生半灌木。中国特有植物。茎半木质化，多分枝。叶 2~3 回全裂，小裂片披针状条形，头状花序半球形，簇生成疏伞状。管状花多数，黄色。主要分布在甘肃、青海等地，生于海拔 1 500 米以上的干旱山坡，适口性中等。

326-1　丝裂亚菊

326-2　丝裂亚菊花序

A perennial semishrub occurring in hill-slopes, distributed in south of Loess Plateau. The feeding value is moderate and livestock like to graze. The palatability is good when it is young and fresh.

327　细叶亚菊

学名　*Ajania tenuifolia*
英名　Thinleaf Ajania

多年生草本。茎直立或斜升,多数自基部分枝,幼嫩时绿色,老熟后逐渐变为灰白色,叶片2~3回分裂,小裂片条形,两面具白色短柔毛。头状花序多数在枝端排列成伞房状,总苞钟状或卵圆形。生于海拔2 000米以上的干旱山坡草地,青绿期家畜很少采食,经霜后牛、羊喜食花序,缺草时家畜采食其茎叶。

A perennial herb, commonly found on sunny mountain side, distributed in south and central Loess Plateau and Hebei, Shandong, etc.The feeding value is moderate.Cattle and sheep like to graze when it is fresh.

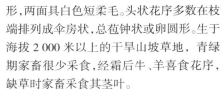

327-1　细叶亚菊　　　　327-2　细叶亚菊花序

328　大花千里光

学名　*Senecio ambraceus* Turcz.
英名　Amber Groundsel

别名琥珀千里光、千里光。多年生草本。茎曲折,多分枝,高达1米以上,近无毛。基生叶倒卵形,大头羽状深裂或不裂。中上部叶片长三角形,顶端长渐尖,基部截形或斧形至心形,边缘有浅裂或深裂齿。头状花序多数,排列成复总状伞房花序。总苞半球形,舌状花黄色,13~15片,管状花多数。瘦果圆柱形,冠毛淡白色。主要分布在陕西、甘肃、河北、河南以及内蒙、东北,在光照和水分条件好的草甸草原生长良好。属于中等牧草,家畜喜食其幼嫩枝叶,开花粗老后可调制青干草。

A perennial herb reproduced from seeds.It is commonly found in grassland of mountain area, distributed in provinces in east, south and southwest China as well as Gansu, Shaanxi.The feeding value is low.Cattle and sheep will graze in winter.

328-1　大花千里光　　　　　　　　328-2　大花千里光花序

329　黄缨菊

学名　*Xanthopappus subacaulis* C.Winkl.
英名　Common Xanthopappus

多年生无茎草本。根状茎粗壮,茎部被多数纤维状的残存叶柄。叶革质,莲座状,叶柄长约5厘米,扁平,具多数沟纹,密被蛛丝状绒毛;叶片长圆状披针形,先端渐尖,具刺尖头,基部延伸成叶柄,羽状深裂,裂片卵形,边缘有不规则小裂片,裂片先端具硬刺,表面绿色,无毛,有光泽背面密被白色蛛丝状绒毛。头状花序大形,无梗或具粗梗具蛛丝状毛,梗上具苞片2,总苞片多层先端刺尖,花冠黄色,长3~4厘米,筒部与檐部近等长,檐部5裂片线形。果实黄色,刚毛状,有极短的羽毛。花粉粒近球形,具3孔沟,内孔模长,椭圆形,生长于宁夏、甘肃、青海等地。

329-2　黄缨菊花序

329-1　黄缨菊

A perennial stem-less herbaceous plant, reproduced from seeds. It occurs on mountain meadows in Qinghai-Xizang plateau and the southwestern part of Loess plateau. Cattle and sheep have limited grazing after autumn frost.

330　聚头蓟

学名　*Cirsium souliei* (Franch) Mattf
英名　Soulie Thistle

多年生草本。无茎或近无茎。叶基生,莲座状,羽状分裂,轮廓长条状,无柄,紧贴地面;叶背密被白色蛛状绵毛,叶面无毛或疏毛,叶缘具刚毛和突出的硬刺。头状花序多个,聚集在叶丛中心,总苞短钟状,苞片多层,管状花红色;瘦果楔形,冠毛多数,白色。分布于青藏高原及其与黄土高原接界地带的草原、草坡或沟壑,枯萎后牛、羊采食。

A perennial herbaceous plant reproduced from seeds. It occurs in alpine meadows or mountain grasslands, and is distributed in Qinghai-Xizang plateau and the parts of the northwest Loess plateau. Cattle and sheep will graze it in autumn.

330　聚头蓟

331 黄帚橐吾

学名 Ligularia virgaurea
英名 Goldenrod Goldenray

多年生直立草本。单生,茎上具纵条棱,无毛,基部通常被残叶片纤维包围。基生叶和下部叶直立,叶柄长5~10厘米,叶片卵形,倒卵状披针形,先端渐尖,急尖或钝圆,基部渐狭楔形下延,两面无毛,近全缘或具疏细齿,叶脉羽状,中脉粗壮,侧脉4~8对,在近边缘处网结;上部叶渐小,无柄,线状披针形,先端渐尖,基部抱茎。头状花序数个至多数,初直立后下垂,在茎端排列成总状;被疏柔毛,具线形苞叶和1~2片小苞叶;总苞宽钟状,总苞片10~12,线状披针形,先端渐尖,边缘膜质,无毛。舌状花9~14,舌片线形,先端

331-1 黄帚橐吾　　331-2 黄帚橐吾花序

具3齿;筒状花14~20,长约8毫米;雄蕊伸出;花柱分枝反卷。主要生长在甘肃、青海。

A perennial herbaceous plant.It is a companion species of beach forb meadows and it also grows on riverside swamps.It is distributed over northern,northwest and southwest China.Usually,animals do not graze.

332 箭叶橐吾

学名 Ligularia sagitta
英名 Schmidt Goldenray

多年生草本。茎直立,上部被绵毛。叶片三角状卵形,具长柄,柄具不规则翅;下部叶急狭,叶柄基部扩大抱茎,叶片戟形或稍心形,顶端钝或有小尖头,边缘有细锯齿;上部叶狭长至条形。头状花序松散的于枝顶排列成总状,花梗长,被蛛丝状毛;有条形苞叶,舌状花5~9朵,舌片黄色,矩圆状条形,瘦果圆柱形。分布于青藏高原及毗邻地带的高山草原,内蒙古高原,是草原退化的标志性植物。

It is a perennial herbaceous plant, commonly found in alpine meadows and river sides.It is distributed in southwest Loess plateau, Qinghai-Xizang plateau and Inner Mongolia plateau.The feeding value is low.

332-2 箭叶橐吾花序　　332-1 箭叶橐吾

333 狗舌草

学名　*Tephroseris kirilowii*
英名　Kirilow's tephroseris

多年生草本。被白色蛛丝状密毛。基生叶与茎下部叶密集，卵形或宽卵形，呈莲座状；茎生叶少数，条状披针形，基部抱茎。头状花序伞房状排裂，总苞背面被蛛丝状毛。舌状花黄色，筒状花多数。瘦果圆柱形，被密毛，冠毛白色。生于山地草坡、林缘。分布我国北方，包括甘肃、陕西、山西等省。低等牧草，牛、羊可以采食。该草含有某种酸性物质，能刺激皮肤发痛，制成干草后则无不良影响。

A perennial herb, mainly reproduced from seeds. It exists among thickets, under forest in hill slopes. It is distributed in northern China, including Gansu, Shaanxi and Shanxi provinces. The feeding value is low and sheep will graze a little. The palatability is better as hay.

333　狗舌草

334 蒙古鸦葱

学名　*Scorzonera mongolica* Maxim.
英名　Mongolian Serpentroot

别名羊角菜。多年生草本。株高6～30厘米，通体灰绿色，无毛。茎多数，直立或自基部铺散。叶质厚，基生叶披针形或条状披针形，基部渐狭成短柄；茎生叶无柄，条状披针形。头状花序单生于枝端，舌状花黄色。瘦果圆柱状，有纵肋，冠毛白色，羽毛状。多生长在轻盐荒地，主要分布于黄土高原及其周边的青海、内蒙古、新疆等地。幼嫩茎叶是优质饲料，马、牛、羊、驴、骡等家畜均喜食。

A perennial herbaceous plant reproduced from seeds. It occurs on light sandy saline soils in arid region, and is distributed in north and northwest China. The feeding value is high and cattle, horses, camel and sheep will graze it.

334-1　蒙古鸦葱

334-2　蒙古鸦葱茎

335 艾 蒿

335-3 艾蒿花序

学名　*Artemisia argyi*
英名　Argy Wormwood

多年生草。根系发达,茎直立,密被白色或银白色柔毛,具特有的清香味。叶互生,叶片羽状深裂或浅裂,上部多为花序枝,少或无裂叶,下部裂片大而多,叶面具白色丝状毛和腺点。头状花序,排列成复总状,总苞卵形,总苞片3~4层,边缘膜质,背部有毛。蒴果矩圆形,以根茎繁殖为主。广布我国北方,喜生于潮湿和肥沃的荒地。适口性较差,初春家畜采食,秋后采食种子、花序和枯叶。

335-1 艾蒿

335-2 艾蒿叶

A perennial herb, spread by crown buds and by seeds. It occurs in wastelands, roadsides and field-sides, distributed in the northern, northeastern and western parts of the country. Horses, cattle and sheep will graze it in the autumn and winter.

336 黄花蒿

学名　*Artemisia annua* L.
英名　Bweet Wormwood

因具臭味,也叫臭蒿。一年生或越年生草本。茎直立,高大,粗壮。上部多分枝,无毛。叶互生,开花后上部叶小而少,下部叶枯萎,中部叶卵形,2~3回羽状深裂,裂片及基部裂片抱茎。头状花序极多数,球形,有短梗,排列成总状或复总状花序,常有条形的苞叶;总苞无毛,总苞片2~3层,花黄色,筒状。瘦果倒卵形或长椭圆形,有细纵棱。分布广泛,常着生于沟底、水边或水分较好的农田、果园。适口性差,枯萎后,羊采食籽、叶。

A biennial or annual herb reproduced from seeds. It occurs in wastelands, farmlands and grasslands, distributed over all parts of the country. Camels will graze it after withered, sheep and goats grazing is limited in autumn.

336-1 黄花蒿　　　　336-2 黄花蒿叶

337 黄 蒿

学名 *Ariemisia scoparia* Waldst
英名 Virgate Wormwood

别名猪毛蒿、茵陈。一年生或二年生草本。茎直立，带紫褐色，有多数开展或斜升的分枝。叶密集，长圆形，2或3回羽状全裂，裂片丝状条形或毛发形，常密被柔毛。头状花序极多，总苞近球形。瘦果长圆形。适生于山坡、沟谷、沙丘和任何空地上。山羊、绵羊都喜食，适口性随生长变老而变差。花期刈割晒制的干草是冬春的好饲料。

A biennial or annual herb reproduced from seeds.It is commonly found in wastelands, farmlands and in young forest growth.It is distributed over all parts of the country.The feeding value is high. Sheep, cattle, goats, camels, horses, donkey and mule like to graze in spring and autumn.

337-2 黄蒿幼苗　　　337-1 黄蒿

338 细裂叶莲蒿

学名 *A.Santolinaefolia* Turcz.
英名 Thinlobed Wormwood

小灌木。多丛生，丛中枝条直立，周围枝条斜升，少分枝，有不均匀的暗红紫色。基部叶片大而多，互生，具柄，有毛，柄具翅，轮廓三角状，2~3回羽状深裂或全裂，小裂片条形，先端尖；叶面淡绿色，背面密被灰白色绵毛。头状花序在枝条中上部密集成总状花序。旱生植物，主要分布于黄土高原中部、北部的山坡、地梗，属于中等牧草，青绿时家畜很少采食，经秋霜枯黄后绵羊、山羊采食。

A semi-shrub, reproduced by crown buds and by seeds. It occurs on south facing low mountains and roadsides.It is distributed in north and southwest China.The feeding value is moderate with sheep and goats willing to graze when fresh.

338-1 细裂叶莲蒿　　　338-2 细裂叶莲蒿叶

339　大籽蒿

学名　*Artemisia sieversiana* Willd.
英名　Sievers Wormwood

339-1　大籽蒿

339-2　大籽蒿茎

339-3　大籽蒿花序

一年生或越年生草本。主根较发达，粗壮。茎直立，坚实，具明显的纵沟棱，被白色短柔毛。基生叶易早枯，茎中下部叶具柄。叶片宽卵形或宽卵圆形，2~3回羽状全裂，侧裂片2~3对，小裂片条形或条状披针形，两面被伏柔毛。上部叶逐渐变小，羽状全裂。最顶端的叶不分裂，小，披针形。头状花序半球形，直径4~7毫米，有梗，下垂。在茎上排列成中度扩展的圆锥形。总苞3~4层，被白色伏毛或无毛，花托凸起，密被毛。边缘小花雌性，中央小花两性，花冠狭圆锥状。瘦果卵形或椭圆形，褐色。主要分布于"三北"和西南，多生于山坡草地、田埂、路旁及沟底。营养价值高，是家畜的优良饲草。

A biennial or annual herb reproduced from seeds.It exists in hill-slops, river banks and wastelands with high water.It is distributed in provinces of north, northwest and northeast China.Cattle and camels like to graze after it is withered and sheep and goats will graze its tender braches and seeds.

340　黑沙蒿

学名　*Artemisia ordosica* Krasch.
英名　Ordos Wormwood

别名油蒿、沙蒿。半灌木，高50~70(100)厘米，主茎不明显，多分枝。老枝外皮暗灰色或暗灰褐色，当年生枝条褐色至黑紫色，具纵条棱。叶稍肉质，1或2回羽状全裂，裂片丝状条形；茎上部叶较短小，3~5全裂或不裂，黄绿色。头状花序多数，卵形，通常直立，具短梗及丝状条形苞叶，枝端排列成开展的圆锥花序；总苞片3~4层，宽卵形，边缘膜质；有花10余个，外层雌性，能育；内层两性不育。瘦果小，长卵形或长椭圆形。多生长在干草原或荒漠化草原，分布于晋、陕北部、甘肃西北和内蒙古。在饲草季节性平衡中有一定意义，骆驼的主要饲草，其他家畜一般情况下不采食。

A semi-shrub that grows on fixed dunes, sands and sand-covered soils.It is distributed over north and northwest Loess plateau.It is the main forage of livestock in the winter and spring.Goats and camels like to graze and can be fed as hay for sheep.

340-2　黑沙蒿花序　　340-1　黑沙蒿

341 蒙古蒿

学名　*Artemisia mongolica* Fisch. et Bess.
英名　Mongol Sagebrush

多年生直立型草本。茎单一，具纵棱，常带紫褐色，被蛛丝状毛。茎生叶在花期枯萎；中部叶具短柄，基部抱茎；羽状深裂叶具3～5深裂的小裂片，边缘有少数锯齿或全缘，顶裂片又常3裂，裂片披针形至条形；叶上面绿色，近无毛，下面密被短茸毛。花序枝斜向上升，头状花序矩圆状钟形，具短梗或无梗，边缘小花雌性，中央小花两性；花冠伏钟形，紫红色。瘦果矩圆形，深褐色，无毛。生于沙地、河谷、撂荒地或路旁，分布在"三北"及西南、华南、华中各地。中等牧草，各种家畜均采食。

341-2　蒙古蒿花序　　　341-1　蒙古蒿

A perennial erect herb, spread by crown buds and by seeds. It exists in sandy lands, riversides, roadsides and wastelands. It is distributed over parts of the north, northwest, northeast, southwest China. The feeding value is moderate and horses, cattle, sheep and goats will graze it in the spring, autumn and winter.

342 茭蒿

学名　*Artemisia giraldii* Pamp
英名　Girald Wormwood

别名吉氏蒿、蒿、灰蒿。半灌木状草本。根粗壮，褐色。茎单一、直立，较粗硬，仅基部木质化，常带红紫色，被灰白色柔毛。基生叶与茎下部叶于花期枯萎；中部叶椭圆形，指状3深裂，裂片狭条形，先端钝或锐尖，两面被伏贴柔毛；上部叶较小，不分裂或3全裂。头状花序宽卵形或矩圆形，多数在茎上部，排列为扩展的圆锥状。总苞片3～4层，无毛，中肋绿色，具宽膜质边缘。边缘小花雌性，中央小花两性，管状。瘦果倒卵形，褐色，甚轻小。主要生长在干旱山坡、陡坡、悬崖处。春季幼嫩时或经秋霜后，放牧家畜皆喜食。

342-2　茭蒿花序　　　342-1　茭蒿

A semi-shrub or perennial herb, it exists in the steep hill, edges of precipice and among thickets in mountain valleys. It is distributed over the Loess plateau. The feeding value is moderate or high, livestock animals like to graze when fresh in spring and sfter frost in autumn.

343 冷 蒿

学名　*Artemisia frigide* Willd.
英名　Frigid Sagebrush

343-1 冷蒿　　343-2 冷蒿茎　　343-3 冷蒿叶

别名小白蒿、串地蒿、兔毛蒿。小半灌木，全株密被灰白色绢毛。根状茎横走。茎多条，丛生，基部木质，斜升或直立。叶具短柄或无柄，2～3回羽状全裂，小裂片披针状条形或条形，基部的裂片抱茎成托叶状；上部叶小，3～5羽状或近掌状全裂；花序枝上的叶不裂，条形。头状花序半球形，具短梗，下垂，多数，在茎上部排列成总状或圆锥状；总苞片3层，透明膜质，中部灰绿色，先端钝；边缘小花雌性，9～12枚，花冠细管状，黄白色，中央小花两性，多数，淡黄色。花托凸起，有托毛。瘦果矩圆形，褐色。温带旱生植物，耐寒。主要分布"三北"及内蒙古。羊、马四季喜食，是该地区极为重要的抓膘、保膘、催乳的优良牧草之一。

A small semi-shrub, it widely grows on steppes and desert steppes. It is distributed over northern Loss plateau and north of China. The feeding value is the highest in the composite family, used for finishing lamb and lactating animals. It is palatable throughout the year.

344 耆状亚菊

学名　*Ajania achilloides*
英名　Yarrow-like Ajania

344-1 耆状亚菊　　344-2 耆状亚菊茎

小半灌木。根木质，粗壮，直伸，老枝短缩，黄褐色，自不定芽发出多数花茎枝；花茎分枝或仅上部具短花序分枝，淡绿色，有纵条棱被贴伏的白色短柔毛。叶2回羽状全裂，1回裂片2对，2回裂片线形，先端钝圆，两面灰绿色，被贴伏的白色短柔毛。叶脉不明显；上部叶和下部叶小，羽状全裂。头状花序小，少数在茎端排列成复伞房花序；梗短。总苞钟状；总苞片4层，黄色有光泽，先端钝圆，边缘白色膜质，无毛，外层卵形或卵状披针形。中内侧的椭圆形至披针形。花冠细筒状，先端4深裂齿；果实倒卵形，褐色。

A semi-shrub, it commonly grows on alkaline or saline-alkali soils in steppes. It is distributed all over the Loess plateau. The feeding value is low and usually sheep and goats limit graze.

五十一 香蒲科

345 宽叶香蒲

学名 *Typha latifolia* L.
英名 Common Cattail

多年生草本。匍匐根状茎发达,多生于水边或浅水中。茎直立,粗壮。叶片条形,叶鞘圆筒状,边缘膜质化。花单性,雌雄同序,花序圆锥状,雄花序居上,雌花序在下,略长于雄花序。雌花序红褐色,如棒状海绵,与雄花序相接。适口性差,只有在幼嫩或干燥时家畜采食。

A perennial paludous herb.It occurs in marshlands, shallow water and paddy fields around plain and low mountain areas.It is distributed in provinces of north, northwest China as well as Sichuan, etc.When it is very fresh, cattle, sheep and pigs like to graze it.

345-3 宽叶香蒲穗(雌)　345-2 宽叶香蒲穗(雄)　345-1 宽叶香蒲

五十二 禾本科

346 老芒麦

学名 *Elgmus sibiricus* L.
英名 Sibirian Wildryegrass

禾本科披碱草属多年生疏丛型草本植物,别名西北利亚披碱草。根系发达,呈须状。茎直立,基部稍倾斜。叶扁平,两边粗糙或下面平滑。叶鞘光滑,位于植株下部的长于节间。叶舌短,膜质。无叶耳。穗状花序,较疏松,略下坠或向外曲展,穗轴分节,一般每节着生2枚小穗。颖狭披针形,具3~5条明显脉纹,顶端渐尖或具短芒。内外稃几等长,外稃背面被短毛,顶端延伸成芒,成熟时向外弯展。颖果长扁圆形,尖端具毛茸,极易脱落。老芒麦抗严寒、喜湿润,北半球广布野生种,我国东北、西北、华北、西藏等地野生和人工栽培均有分布。因其叶量大(抽穗期茎叶比为1:0.6)草质柔嫩,适口性好。放牧、青饲、青贮、调制青干草,均为马、牛、羊、兔等家畜所喜食。

A perennial grass reproduced from seeds.It grows on hillsides, hills and mountain forest edges.It is distributed over northern and southwestern China.Feeding value is high.Its texture is softer than Wildryegrass's and the palatability is good for cattle and sheep.

346-1 老芒麦

346-2 老芒麦花穗

347　垂穗披碱草

学名　*Elymus nutans* (Griseb) Nevski
英名　Drooping Wildryegrass

多年生疏丛型草本植物。根系发达，呈须状。茎直立，3～4节，茎部稍呈膝曲。叶片扁平，两面粗糙或上面光滑。叶鞘除基部者外，其余均短于节间。穗状花序，较紧密，小穗排列少偏于一侧，通常弯曲先端下垂，穗轴每节一般有2枚小穗。接近顶端处各节仅具1枚小穗，基部具不育小穗，小穗具短柄或无，小穗绿色，成熟后变紫。颖长圆形，先端钝圆或截平。颖果，种子披针形，紫褐色。垂穗披碱草原为野生种，我国西藏、西北、华北等地都有分布。现高海拔冷凉牧区多有栽培，其垂直分布的上限可达4 200米，有异乎寻常的适应能力。该草叶量较少，适时放牧或刈割调制，马、牛、羊均喜食。

A perennial grass reproduced from seeds. It is commonly found on the forest edges, meadows and roadsides, and is distributed over northern and southern China. It is an excellent forage with high nutritive value for cattle, sheep and horses.

347-1　垂穗披碱草　　347-2　垂穗披碱草花序

348　无芒雀麦

学名　*Bromus inermis* Leyss.
英名　Smooth Bromegrass

别名光雀麦、禾萱草、无芒麦。禾本科多年生根茎型草本植物。须根发达具地下走茎，根茎可生出大量须根。茎4～6节，直立，圆形，粗壮。叶披针形，淡绿色，叶缘具短刺毛。叶鞘圆形，闭合，紧抱茎，叶舌膜质，叶无耳。茎、叶、节均光滑无毛。圆锥花序，开展，分枝细，穗轴每节上轮生穗枝梗2～3个，2片颖均为披针形，大小不等，膜质，不易脱落。外稃边缘膜质，无芒或具短芒。种子扁平，呈艇形。返青早，枯黄迟，生长期长，耐践踏，宜放牧，再生性好，可于苜蓿混播，营养丰富，适口性极好，各种家畜均喜食。

A perennial grass that is spread by rhizome and by seeds. It is commonly found on meadows and steppe and it is suitable for cultivation. It is distributed over northern China. The feeding value is high and it is one of the best forages in the world. The texture of the grass is soft, with large amount of leaves. Palatability is good.

348-2　无芒雀麦花序　　348-1　无芒雀麦

349　多节雀麦

学名　*Bromus plurinodis* Keng
英名　Manynode Bromegrass

多年生草本。须根稀疏，较坚韧，茎秆直立，常单生，高达 100 厘米以上，具节。叶鞘均长于节间，叶舌膜质，褐色，常撕裂。叶片扁平，边缘粗糙，上面具白色柔毛。圆锥花序，每节分枝 2~4 枚，斜向上升。颖披针形，边缘膜质；具细而劲直的芒，背面下部边缘及脉处均具微毛。常生于针叶林边缘的灌丛之中，喜湿润阴凉环境。叶量丰富，草质柔软，青鲜时，马、牛、羊均喜食。

A perennial grass, it spread by rhizome and by seeds. It occurs on meadows, forest fringes and mountain valleys. It is distributed in northern China and Qinghai-Xizang plateau. The feeding value is high, and it is one of the excellent forage.

349-2　多节雀麦花序　　　349-1　多节雀麦

350　冰　草

学名　*Agropyron cristatum* (L.) Gaertn
英名　Wheatgrass, Crested Wheatgrass

多年生疏丛型草本植物。根系发达，须根密集，具沙套，有时有短根茎。茎秆直立，2~3 节，基部节呈膝曲状，上披短绒毛。叶披针形，叶背光滑叶面密生绒毛。叶鞘短于节间，紧抱茎，叶舌不明显。穗状花序，有小穗 30~50 个。小穗无柄，紧密排列于穗轴两侧，呈蓖齿状。颖不对称，沿龙骨上有纤毛，外颖尖端芒状，外稃有毛，顶端具短芒。冰草系全世界温带地区主要牧草之一，我国东北、西北、内蒙古等地区均有栽培。该草茎叶嫩脆，粗蛋白质含量高，营养丰富，适口性好，马、牛、羊均喜食。

350-1　冰草

A perennial grass that is usually cultivated in arid regions of the Loess Plateau. It germinates in early spring and can tolerate heavy grazing. The feeding value is high. It is palatable for horses and cattle when young, while sheep and goats grazing is limited.

350-2　冰草花序

351 虉草

学名　*Phalaris arundinacea* L.
英名　Reed Canarygrass

别名草芦、园草芦。多年生草本。根系入土可达 2.5～3.0 米。茎秆直立，光滑无毛，似有蜡质。叶片平展，叶稍较节间为长，上部较短，光滑；叶舌质薄；无叶耳。圆锥花序，密而狭，开花时分枝展开，花后收紧；小穗丛密，每小穗具 3 花，其中 2 个花不孕。种子长椭圆形，光滑，灰黑色。在庆阳黄土高原试验站种植结果表明，对土壤要求不严，高水肥条件下生长发育快，能高产。叶量大，质地柔软，适口性好，各种家畜喜食。

351-1 虉草　　　351-2 虉草花序

A perennial grass reproduced from seeds. It is usually suited to fertile and wet lands and the yield is high. The texture of the grass is soft, with large amount of leaves, palatability is good. Livestock animals will graze.

352 硬质早熟禾

学名　*Poa sphondylodes* Trin. ex Bunge.
英名　Hard Bluegrass

别名铁丝草。多年生草本，种子繁殖。茎秆丛生，直立，细而硬，高 30～60 厘米，具 3～4 节，顶节裸露的部分很长。叶鞘无毛，叶舌膜质，先端尖锐；叶片狭窄。圆锥花序紧缩，分枝短而粗糙；小穗含 4～6 花，颖果纺锤形。生于典型的草原地带，也可以进入林间灌丛，园圃、路边，广布"三北"地区。幼嫩时各种家畜皆喜食，以马、羊为最。

A perennial grass reproduced from seeds. It occurs mainly in grasslands, roadsides and hill-slopes. It is distributed in north, northeast and northwest China as well as Shandong and Jiangsu. Livestock will graze it when it is fresh.

352-1 硬质早熟禾　　　352-2 硬质早熟禾花序

353 高羊茅

学名　*Festuca arundinacea* Schreb
英名　Tall Fescue

别名苇状羊茅、苇状狐茅，原产西欧。多年生草本植物。须根发达，入土较深。茎直立，分4~5节，疏丛型。叶带状，叶背光滑，叶面粗糙。基生叶密集丛生，叶量丰富。圆锥花序，松散多枝，每小穗含4~5朵小花，颖窄披针形，外稃无芒或具短芒。颖果棕褐色。调制青干草或放牧，所有草食家畜均喜食。

A perennial grass reproduced from seeds. It is a good plant for artificial grassland. It is one of the excellent forage in the world. The feeding value is high. It is palatable in spring, autumn and winter.

353-1　高羊茅

353-2　高羊茅花序

354 紫羊茅

学名　*Festuca rubra* L.
英名　Red Fescue

别名红狐茅，原产欧亚大陆和北非。羊茅属多年生草本植物，具纤细的须根和短根茎。疏丛型，茎直立或斜生，基生叶纤细密集，形成低矮的下繁草。叶片对折或内卷成线形。叶背光滑，叶面有绒毛。绿色或深绿色。圆锥花序，窄长稍下垂，分枝较少，开花时散开。小穗含小花3~6朵。外稃披针形，具小短芒。我国"三北"地区及西南均有野生种和栽培种，草质优良，家畜喜食。

A perennial grass reproduced by seed or by rhizome. It is suited to fertile and wet lands, and it is distributed over northern China. The feeding value is high and the texture of the grass is soft. When it is fresh, sheep, goats and cattle are willing to graze, and it can be used as hay.

354-2　紫羊茅

354-1　紫羊茅花序

355 猫尾草

学名 *Phleum pratense*
英名 Timothy

别名梯牧草。多年生草本植物。须根系，疏丛型，茎基部有球状短根茎。茎秆直立。叶片扁平，叶鞘松弛，着生于节间。圆锥花序呈圆柱状，小穗长圆形，含1朵小花。颖果细小、圆形。原产欧亚大陆，我国"三北"地区也有种植。喜冷凉湿润气候，宜刈割调制干草，是大家畜（马、牛、驴、骡、骆驼等）的良好饲料。

A perennial grass reproduced from seeds. It grows in cold and wet region of the Loess Plateau. It is a good forage for artificial grassland. The feeding value is moderate. Horse, donkey and mule like to graze, and cattle grazing is limited, but sheep do not like it.

355-2 猫尾草花序

355-1 猫尾草

356 鸭茅

学名 *Dactylis glomerata* L.
英名 Common Orchardgrass

别名鸡脚草、果园草。多年生草本植物。须根系，茎直立或基部膝曲，疏丛型。叶片蓝绿色，幼叶为折叠状。基部叶片密集下披。圆锥花序，开展，小穗聚集于分枝的上端，含2~5朵小花。种子为颖果，黄褐色，长卵圆形。原产欧洲、北非和亚洲的温带，我国新疆、四川、云南等地有野生分布，湖南、湖北、江苏等省有大面积的栽培。草质柔软，叶量丰富，适口性好，草食家畜均喜食。

A perennial grass reproduced from seeds. It is a shade plant, and suited in fertile and wet lands. It is commonly found under forest and orchard. The texture of the grass is soft with a large amount of leaves. The feeding value is high and grazed all year around.

356-1 鸭茅　　　　356-2 鸭茅花序

357 燕 麦

学名 *Avena sativa* L.
英名 Tall Oatgrass

别名铃铛草。一年生草本植物。根须状。茎直立,中空,具4~7节。叶片平展,无叶耳。叶舌较大,顶端具稀疏裂齿。圆锥花序,开散顶生。穗轴多直立或下垂,每穗具4~9节,节部分枝。小穗30~50个,每小穗含1~2朵小花。小穗轴不易脱节,颖较宽大,呈薄膜状。外稃具芒,芒出自背脊的中部,也有无芒者。内外稃紧包种粒,不易分开。颖果,纺锤形,狭长,具簇毛,有纵沟。系饲草、饲料兼用作物。是马、骡、驴传统的优质饲料,牛、羊均喜食,早期刈割,饲喂鸡、兔、猪效果奇佳。

An annual grass reproduced from seeds.It is cultivated in north and northwestern Loess Plateau.Usually sowing in May or June and harvested at end of autumn.The texture of the grass is green and soft,palatability is good and nutritive value is high.It is an excellent forage for young animals throughout the winter.

357 燕麦

358 野 燕 麦

学名 *Avena fatua* L.
英名 Wild Oat

一年生或越年生草本。茎秆丛生或单生,直立,具节明显。叶鞘松弛,光滑或基部被柔毛。叶舌膜质透明,叶片宽条形。圆锥花序开展,松散呈塔形,分枝轮生,疏生小穗。小穗含2~3个小花,花梗长,弯曲下垂。两颖片近等长,外稃坚硬,下部散生粗毛,芒从稃体中部稍下处伸出,膝曲,扭转。颖果长圆形,被棕色柔毛,腹面具纵沟。靠种子繁殖,饲用价值高,适口性好,家畜喜食。

An annual or biennial grass reproduced from seeds.It is distributed over all parts of the country,and is a harmful weed in winter wheat.This is a very good forage for horses,cattle,sheep and goats in autumn and winter.

358-2 野燕麦花序

358-1 野燕麦

359 羊 茅

学名　*Festuca ovina*
英名　Thin Fescue

多年生草本，种子繁殖。密丛型，须根状。秆细弱、直立，高20～60厘米，基部有2节，通体无毛。叶片稍内卷，或呈针状，基叶丛生，长而发达。小穗有2至多数小花，排成狭窄或开展的圆锥花序；颖不等长，第一颖常具1脉，第二颖常有3脉；外稃有5脉，短失至有芒；果分离或与稃黏合。适应性广，在西北地区多用于草坪草种。较羊茅属其他几种牧草适口性好，舍饲牛、羊喜食，是优良牧草之一。

A perennial grass reproduced from seeds and by rhizome. It occurs in north, northwest and northeast China. The feeding value is high, the palatability is good and all livestock will graze it when fresh or dry.

359-1　羊茅

359-2　羊毛花序

360 多年生黑麦草

学名　*Lolium perenne* L.
英名　Perennial Ryegrass

别名宿根黑麦草、牧场黑麦草。多年生草本植物。须根发达。茎直立，光滑中空，色浅绿，多分蘖。叶片深绿有光泽，多下披。叶鞘长于或等于节间，紧抱茎。叶舌膜质。穗状花序，每穗有小穗15～25个，小穗无柄，紧密互生于穗轴两侧。有花5～11枚，结实3～5粒。第一颖常常退化，第二颖质地坚硬，外稃质薄，端钝，无芒。内稃和外稃等长，顶端尖锐，透明，边缘有绒毛，颖果梭形。主要分布于华东、华中和西南该草茎叶柔嫩，叶量丰富，草质优良，适口性好。是马、牛、兔、羊、鱼的理想饲料。

A perennial grass reproduced from seeds, it is distributed in south and central China as well as some northern area. It is an excellent forage and germinates in early spring, leaves are soft, feeding value is high. Grazing livestock like to this plant.

360-1　多年生黑麦草　　360-2　多年生黑麦草花序

361 黑 麦

学名 *Secale cereale* L.
英名 Rye

别名洋麦。一年生草本。秆直立,株高 70~150 厘米。叶片披针形,扁平,叶鞘无毛,叶舌近膜质。穗状花序顶生,紧密;小穗含 2~3 小花,下部小花结实,顶生小花不育,颖果。黑麦喜冷凉气候,我国"三北"地区均有种植。黑麦叶量大,茎秆柔软,营养丰富,适口性好,是牛、羊、马的优质饲草,籽粒也是上等的精饲料。

An annual grass reproduced from seeds. It is suited to fields in mountain areas, and is distributed in northwest China and Gansu province. The yield is very high. Cattle, sheep and horses will consume it as hay. In addition, the seeds are good concentrates.

361 黑麦

362 早熟禾(瓦巴斯)

学名 *Poa pratensis* L.
英名 Kentucky Bluegrass

别名六月禾、蓝草。多年生草本植物,须根系,有根状茎。茎直立或斜生。叶窄条形,光滑,浅绿到深绿色。基生叶密集,常形成厚密的草丛。圆锥花序,由 7~9 节组成,每节 3~4 个分枝,小枝上有 2~4 个小穗,每穗 2~4 朵小花。种子为颖果,纺锤形,长约 2 毫米。茎叶柔软,营养丰富,适口性好,是优良的放牧型牧草。因其青绿色长,可建植多种草坪绿地。

A perennial grass reproduced from seeds. It is commonly found in wet grasslands, roadsides and other wetlands, and distributed in most of the country. It is a very good forage for sheep, goats and cattle all year around.

362 早熟禾

363　一年生早熟禾

学名　*Poa annua* L.
英名　Annual Bluegrass

一年生或越年生草本。茎秆丛生，直立或斜生，细弱。叶鞘多自基部以下闭合，无毛。叶舌膜质，圆头，叶片质地柔软。圆锥花序开展，每节分枝 1~2，小穗含 3~5 花，颖膜质，第一颖略短，具 1 脉；第二颖具 3 脉。外稃边缘及顶端膜质，具 5 脉，脊部具长柔毛。颖果纺锤形，先端渐尖，黄白色。果园、农田、菜地及沟底阴湿地方较多，属优良牧草，放牧家畜尤为喜食。

An anuul or biennial grass reprodusdused from seed and by rhizome.It is suited to wet lands in gully, diches and forest edges, distributed in south estern Loess Plateau.It is an excellent forage and all grazing livestock like this plant.

363-2　一年生早熟禾花序

363-1　一年生早熟禾

364　高山早熟禾

学名　*Poa alpine* L.
英名　Alpine Bluegrass

多年密丛型下繁草。根系和细根茎发达，密集于 30 厘米内的土表层，丛生。茎秆高 10~30 厘米，通体光滑无毛。叶片"V"形折或扁平，条形，先端渐尖。圆锥花序，松散开放，呈淡紫色。生于海拔 2 200~3 200 米的山地草甸、高山草甸和沼泽草甸，主要分布在青藏高原及临近地区。高寒牧区夏季草场的优良牧草，5 月上旬方能返青，再生力强，耐踏、耐牧。茎秆细弱，草质柔嫩，营养物质含量丰富，各种家畜均喜食，以鹿为最。

A perennial grass reproduced from seeds and by crown buds.It occurs on meadows, meadow steppes and mountain forest fringes, and is distributed in whole northern China.It is an excellent forage and has good palatability.

364-1　高山早熟禾

364-2　高山早熟禾花序

365 臭 草

学名　*Melica scabrosa* Trin.
英名　Rough Melic

别名枪草、肥马草。多年生草本。茎秆丛生，基部直立或膝曲。叶鞘闭合，光滑，下部叶鞘长于节间，上部叶鞘短于节间。叶片线形，无毛或叶面疏生柔毛。叶舌膜质，先端撕裂而两侧下延。圆锥花序狭窄，分枝紧贴主轴，直立或斜生，小穗柄短，含2~4朵可育的花，顶端有数朵不育的花结成小球体。2颖近等长，膜质，背部中脉具纤毛。外稃长于内稃，且背部粗糙，有7条隆起的脉。颖果纺锤形，褐色。属于优良牧草，宜牧宜刈，家畜喜食。

365-2　臭草花序　　　　　365-1　臭草

A perennial grass reproduced from seeds. It is a common grass found in wastelands, gardens and field ridges, distributed in north and northwest China as well as east and south China, such as Shandong, Jiangsu and Sichuan, etc. when it is fresh, animal do not graze but would graze in winter.

366 赖 草

学名　*Aneurolepidium dasystachys* Nevski
英名　Common Leymus, Common Aneorolepidium

禾本科多年生草本，具下伸的根状茎。秆直立，较粗硬，单生或为疏丛状，茎部叶鞘残留呈纤维状。叶片长10~30厘米，平展。穗状花序直立，颖锥形，小穗先端有短芒。赖草属中旱生植物，分布极为广泛，田边、地埂、荒坡、撂荒地随处可见。春、秋两季家畜喜食，夏季因粗老而适口性降低。

A perennial grass, it is a xerophilous plant, propagated mainly by rhizome. It grows on sands, hills, field ridges and roadsides in the whole Loess Plateau. The feeding value is quite high. It is palatable for cattle and horses when young while sheep and goats grazing are limited.

366-3　赖草根系

366-2　赖草花序　　　366-1　赖草

367 白羊草

367-1 白羊草　　367-2 白羊草花序

学名　*Bothriochloa ischaemum* Keng.
英名　Digitate Goldenbeard

别名白草、茎草、盘棋。多年生草本。短根茎状,疏丛型。茎秆直立或基部膝曲,具3至多节。叶片狭条形,两面疏生疣毛或背面无毛。总状花序多数簇生于茎顶端,小穗成对生于各节。无柄小穗上部有膝曲的短芒,有柄小穗无芒不孕。白羊草是一种广布优势建群植物。返青略晚,夏季生长缓慢,秋季遇雨则迅速生长,入侵和再生能力及强。多生于荒坡、田埂。营养丰富,消化率高,适口性以绵羊最好,其次是兔、山羊、牛、驴、马等。

A perennial grass, spreading by rhizome and by seeds. It is commonly found in hill-slope grasslands, field ridges, wastelands and riverbed. It is distributed in east, south and central Loess Plateau. The feeding value is high, sheep, goats and cattle like this plant all year around.

368 虎尾草

学名　*Chloris virgata* Swartz.
英名　Showy Chloris

别名棒槌草。一年生草本。根须状,株高 20~50 厘米,直立或基部膝曲,丛生,稍扁,光滑无毛。叶片条状披针形,时有卷折,平滑或上面和边缘粗涩,叶鞘松弛,上部叶鞘膨大有时带紫色,最上部叶鞘通常包被花序。穗状花序 4~10 枚簇生于茎顶,初生时如棒槌状,后来逐渐松散呈指状。小穗在穗轴的一侧,呈复瓦状排列两行,每小穗含 2 花,上部花退化不育,并相互抱成球形,下部花两性。第二颖具短芒,内稃稍短于外稃,外稃顶端以下生芒,第一外稃具 3 脉,边脉有长柔毛。多生于干燥山坡、草地,田埂、路边多见。为优良牧草,因草质柔软,无论鲜草,还是调制成干草,各种家畜均喜食。

An annual grass reproduced from seeds. It is commonly found in fields, wastelands and a region of summer-rain in the desert steppes. It is distributed over Loess Plateau. It is an alkali-tolerant plant and animals like to graze.

368-2 虎尾草花序　　368-1 虎尾草

369 中间偃麦草

学名　*Elytrigia intermedia* (Hest) Nevski
英名　Intermediate Elytrigia

别名天兰冰草、中间冰草。多年生草本植物。须根系,具有发达的横走根茎。茎直立,粗壮,疏丛生。基生叶密集,长而宽,茎上叶较多,叶缘有短毛,依此可与其他偃麦草相区别,叶色灰绿,有蜡质。穗状花序,细长。小穗无柄、互生,稀疏斜上排列在穗轴两侧,每穗着生 20~30 个小穗,每小穗有花 3~6 朵。可结实 2~6 粒。原产东欧,我国北方也有种植。可调制干草,可放牧,是草食家畜的好饲料。

A perennial grass reproduced from seeds.It grows wildly and can be cultivated all over the country.The feeding value is high and livestock like to graze as fresh or hay.

369-1　中间偃麦草　　369-2　中间偃麦草花序

370 高冰草

学名　*Thinopyrum Ponticum*
英名　Tall Wheatgrass

多年生密丛型草本。茎秆高度可达 1~2 米,直立,光滑,微被白粉。基叶丛生,茎叶宽长,与茎秆垂直生长。穗状花序较长,小穗稀疏,在穗轴两侧交互着生。小穗上种子排列成纺锤状。耐寒冷、耐水淹、耐盐碱,不耐干旱。原产澳大利亚,20 世纪 80 年代初引入我国,目前东北、西北有一定的栽培面积。不论是青草还是干草,含有特殊的清香味,叶量多,适口性好,各种家畜均喜食,也宜喂家禽和鱼类。可鲜喂,青贮,微贮,也可调制干草。

An erect perennial grass mainly reproduced from seed, it can be resistant to cold, alkali and flooding conditions.It is cultivated wildly and germinates in early spring.It is an excellent forage with good palatability. Livestock animals like this plant as fresh, hay, powder or silage.

370-2　高冰草花序　　370-1　高冰草

371 芨芨草

学名 *Achnatherum splendens* (Trin.) Nevski
英名 Lovely Achnatherum

别名积机草、席萁草。多年生草本。须根粗壮，具沙套，入土深，根幅广，抗逆性强。密丛型，茎直立、坚硬，株高50～250厘米。圆锥花序，颖膜质，披针形或椭圆形，芒直立或弯曲，易脱落。种子、分根均可繁殖。芨芨草具有广泛的生态可塑性。无论春季的嫩枝叶还是秋后的干草，牛、羊、驼、马均采食。

It is a perennial grass with dense tuff stems, salt tolerant. It is constructive species of saline meadows and grows throughout the region. It is distributed over north and northwestern Loess Plateau. The feeding value is high with camels and goats readily grazing while horses and sheep grazing is limited. All animals can use this plant as winter feed.

371-2 芨芨草花序　　371-1 芨芨草

372 柳枝稷

学名 *Panicum virgatum*
英名 Versatile switch grass

多年生草本。丛生型禾草，根茎和种子均可繁殖，根系发达，易形成坚硬的结皮。株高1~2米，直立或斜升。叶片条形，边缘具齿，被疏毛，先端尖锐。穗状圆锥花序顶生，稀疏，披散。庆阳黄土高原试验站于2001年引进栽培，适宜北方和黄土高原生长，具有喜水肥，耐旱、耐水淹，第一年苗期生长缓慢，返青早，生长快，产量高等特点。作为牧草，无论干、鲜，家畜均喜食。适口性不及禾本科牧草无芒雀麦、鸭茅、黑麦草、高羊茅等。

A perennial grass reproduced from seeds and by rhizomome, it can resist cold, drought and alkali soil conditions, but high in yield in fertile lands. The feeding value is moderate. Cattle and sheep will graze its tender branches and seeds.

372-1 柳枝稷

372-2 柳枝稷花序

373 野 稷

学名 *Panicum miliaceum* L.
英名 Wild millet

别名野糜子。一年生草本,种子繁殖。茎秆疏丛生,直立或基部膝曲,高60～120厘米,较粗壮,扁圆形,暴露在叶鞘外的部分密生长疣毛。叶片条状披针形,两面疏生长疣毛;叶鞘短于节间,密生疣毛;叶舌具小纤毛。圆锥花序疏而展,直立,穗轴与分枝有角棱,棱上有毛;小穗长椭圆形;种子椭圆形,成熟后黑色,有光泽。多生于农田和路旁,特别是混杂于糜、谷等作物地中,北方各省均有分布。优质饲草,骡、马等大家畜尤其喜食。

373-1 野稷

An annual grass reproduced from seeds.It occurs in farmlands and wastelands.It is distributed over various regions in northern part of China.The feeding value is high and livestock will graze it as fresh or dry.

373-2 野稷花序

374 湖南稷子

学名 *Echinochloa crusgalii* (L.) var. *frumentacea*
英名 Japannese millet

别名稷子、家稗。一年生草本,种子繁殖。根系发达,根幅宽广。茎直立或基部膝曲而略斜升,粗壮。叶鞘疏松,光滑,无毛,多数短于节间,叶舌叶耳缺,叶片宽条形,边缘具细齿或微呈波状。圆锥花序直立,或成熟时稍向下弯,主轴粗壮具条棱,棱边粗糙至被短硬毛或疏生刺状疣毛,分枝密集,小穗密集排列于小穗轴的一侧,单生或不规则簇生,初为黄绿色,成熟后呈黄褐色。谷粒椭圆形或宽卵形。为高产优质的饲料作物。

An annual grass reproduced from seeds. It mainly grows in southern China, and suited to Loess plateau also.It has high yield and palatability is good.Cattle,horses,sheep and goats will graze it as hay or silage.

374-1 湖南稷子　　374-2 湖南稷子花序

375　升马唐

学名　*Digitaria adscendens*
英名　Ascendent Crabgrass

别名毛马唐、熟地草、拌根草。一年生草本。茎秆基部倾斜或横卧地面,节着地后易生根。叶鞘短于节间,叶片条状披针形,无毛或叶背面被疏柔毛,叶舌钝圆。总状花序3～8枚,呈指状排列于枝顶。小穗披针形,通常孪生于穗轴的一侧,一具长柄,一具短柄。第一颖小,无脉。第二颖长于小穗,有3脉,第一外稃边与小穗等,脉间贴生柔毛,边缘具明显的纤毛。种子厚纸质,顶端尖,背部隆起,内稃薄而紧贴边缘。一般生于农作物下面或多年生牧草地。草质柔软,适口性好,属于优质牧草,牛、羊等家畜乐食。

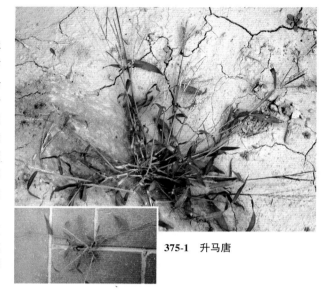

375-1　升马唐

An annual herb reproduced from seeds. It occurs in farmlands, wastelands and roadsides, and mainly distributed in south and central Loess Plateau. The texture of the grass is soft, palatability is good and nutritive value is high. Cattle and sheep like to graze.

376　止血马唐

学名　*Digitaria ischaemum*
英名　Smooth Crabgrass

一年生草本。茎秆基部直立或倾斜,常膝曲。叶鞘疏松裹茎,具脊,略带紫色,无毛或疏被软柔毛。鞘口长柔毛,叶舌干膜质,叶片扁平,披针形,略短于升马唐。和升马唐的主要区别在于总状花序2～4枚呈指状排列,彼此接近,甚至最下面的1枚较离开,小穗在每节着生2～3枚,灰绿色或带紫色。全国普遍分布,晋、豫、甘、陕较多。草质柔嫩,无特殊气味,适口性好,除马以外,其他草食家畜均喜食。

An annual herb reproduced from seeds. It is commonly found in river banks, field sides and damp environments and is distributed over the Loess Plateau. The feeding value is high and cattle and sheep like to graze after an autumn frost.

376-1　止血马唐

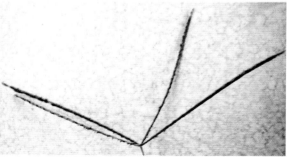

376-2　止血马唐花序

377 狗牙根

学名　*Cynodon dactylon*
英名　Dogtoothgrass

别名铁线草、绊根草、爬地草。多年生草本，种子与营养两种繁殖。具根状茎及匍匐茎，节间长短不等。秆匍匐地面，长可达 2 米。叶舌短，具小纤毛；叶片条形。穗状花序 3~6 枚，呈指状排列于茎顶；小穗排列于穗轴的一侧，含 1 小花；颖片等长，1 脉成脊，短于外稃；外稃具 3 脉，脊上有毛；内稃约与外稃等长，具 2 脊。分布于黄河以南各省，耐放牧，耐践踏；在华北、西北也有生长。叶量丰富，草质柔软，马、牛、羊、兔均喜食，幼嫩时猪和家禽也采食。

A perennial herb propagated mainly by stolon. It exists on fertile and wet lands. It is distributed in provinces and autonomous regions south to the Yellow River valley. The feeding value is high. Horses, cattle, sheep and rabbits like to graze. Pig and poultry will graze its tender branches and leaves.

377-1 狗牙根茎　　　377-2 狗牙根

378 牛筋草

学名　*Eleusine indica*
英名　Goosegrass

别名蟋蟀草。一年生草本。茎秆丛生，从中心近直立或斜升，周围几乎平贴地面。叶鞘扁而具脊，鞘口有柔毛，叶舌短，叶片条形。穗状花序 2~7 枚，呈指状排列于茎秆顶端，有时其中的一两枚着生于其花序的下方。小穗双行密集于穗轴的一侧，含 3~6 朵小花。颖和稃均无芒，内稃短于外稃，脊上具小纤毛。囊果卵形，有明显的波状皱纹。我国各地均有分布，草质柔软，适口性好，多种家畜乐于采食。

An annual herb reproduced from seeds. It is commonly found in rather moist farmlands and roadsides, distributed over all parts of the country. It is a forage with moderate feeding value. The leaves are soft and they are good feed for cattle and sheep.

378-1 牛筋草　　　378-2 牛筋草花序

379 白 茅

学名 *Imperata cylindrica*
英名 Lalang Grass

多年生草本。匍匐根状茎横走,极为发达,黄白色。茎秆直立或斜生,有2~3节,节具长柔毛。老熟时基部常丝裂成纤维状。叶片条形或条状披针形,叶背主脉明显突出;叶舌膜质,头钝。圆锥花序圆柱状,分枝短而密集。小穗披针形或长圆形,基部密生丝状白色长柔毛,柔毛长而且遮盖小穗。以根茎繁殖为主,节上长出新的苗和根系,在农田中繁殖极为迅速。一种十分优良的牧草,家畜喜食,特别是牛和羊。

It is a perennial grass, spread by rhizome and by seeds, and commonly found in hill-slopes, meadows, farmlands and roadsides. It is distributed throughout the country. The feeding value is high and cattle, sheep, goats, and other livestock are readily to graze it.

379-1 白茅草　　379-2 白茅草根茎

380 无芒稗

学名 *Echinochloa crusgalli* (L.) Beauv.
英名 Beardless Barnyardgrass

一年生草本,种子繁殖。茎秆丛生,扁平,光滑,基部斜升或膝曲。圆锥花序直立或下垂,上部紧密,下部稍松散,绿色。小穗密集于穗轴的一侧,有硬疣毛。本品种与稗相似,主要区别是花序有小的分枝,小穗无芒,即使有芒也不超过3毫米。优良牧草之一,叶量丰富,草质柔软,适口性好,马、牛、羊最喜食。

An annual grass reproduced from seeds. It occurs in low lying damp farmlands or wet water sides, and is distributed over all parts of the country. It is an excellent forage, cattle, horses, sheep and goats will graze it anytime.

380-1 无芒稗　　380-2 无芒稗花序

381 长芒野稗

学名　*Echinochloa crusgalli* L
英名　Barnyardgrass

一年生草本。茎秆苗壮丛生，直立。叶条形，无毛，先端渐尖，两侧边缘多变得肥厚，遇干旱时叶片容易内卷；叶鞘光滑无毛，无叶舌。圆锥花序较狭窄，柔软下弯。小穗宽卵形至卵圆形，淡绿色或带紫色，毛较少，具芒，一般芒长1~2厘米。种子成熟后易脱落，落地后繁殖能力强，也极易混入粮、油及其他作物种子中，有时在肥沃农田和季节性休闲地繁殖生长旺盛，该草鲜嫩多质，家畜喜食，调制成干草，品质亦十分优良。

An annual grass reproduced from seeds. It grows in shallow water and wet lands, and is also a common barnyard grass in ditches and rice fields. It is mainly distributed in south of China, and commonly found on fertile and wet lands in Loess plateau. It is an excellent forage.

381-2　长芒野稗花序　　381-1　长芒野稗

382 西来稗

382-1　西来稗　　382-2　西来稗花序

学名　*Echinochloa crusgalli*
英名　Alkali Barnyardgrass

一年生草本。茎直立或斜升，叶披针状至狭线形，叶缘变厚而粗糙，叶舌仅存痕迹。圆锥花序尖塔形，直立或微弯，分枝单纯，无小分枝。小穗无芒，长卵圆形，先端具长尖头。颖及第一外稃具较少的毛，脉上具长毛，无瘤基。颖果椭圆形，凸面有纵脊，黄褐色。与旱稗的主要区别在于花序上部紧凑，花序分枝贴主轴，小穗无芒。

An annual grass reproduced from seeds. It is a serious weed in rice crop, and distributed in south of the Loess Plateau and southeastern China as well as some parts of Sichuan, Hebei, etc. Cattle, sheep and goats would graze its fresh leaves and seeds.

383 囡草

学名　*Beckmannia syzigachne*
英名　American Sloughgrass

一年生草本,种子繁殖。疏丛型,茎秆直立,基部节微膝曲,光滑无毛。叶鞘长于节间,叶舌透明,膜质;叶片扁平,两面粗糙。穗状花序,有短柄,着生于茎顶聚合成为圆锥花序;内、外颖半圆形,泡状膨大,背面弯曲,稍革质。喜生于水湿地、河湖岸旁、浅水中,也是稻田中的杂草。分布于"三北"及西南、华东。早期生长快,枝叶繁茂,家畜喜食,适宜于调制青干草。

An annual grass reproduced from seeds. It occurs in paddy fields, damp grasslands and water sides. It is distributed in north, east and southwest China. The feeding value is high, cattle and sheep will graze it when fresh or dry.

383-1　囡草　　　　383-2　囡草花序

384 大画眉草

学名　*Eragrostis cilianensis* (All.) Link
英名　Stinkgrass

一年生草本。多连片丛生。茎柔弱,自基部开放性向上斜升,节下常有一圈腺体。叶片条形,边缘具腺体。叶鞘短于节间,脉上有腺体,鞘口有柔毛,叶舌退化为一圈短毛。圆锥花序长圆形或塔形,分枝较多而韧,单生,小枝和小穗柄上均有腺体。小穗长圆形,灰绿色至乳白色,含多个小花,两颖近等长,脊上也有腺体。大画眉新鲜时有一种较为特殊的气味。颖果球形,靠种子繁殖。广布于我国各地,多生于草地、撂荒地、各种农田,属优质牧草,家畜喜食。

An annual grass reproduced from seeds. It is commonly found in farmlands, worn-lands and most popular in sandy lands, and distributed over all parts of the China. The feeding value is moderate. Cattle, sheep and goats like to graze after a frost in autumn.

384-1　大画眉草　　　　384-2　大画眉草花序

385 小画眉草

学名 *Eragrostis poaeoides* Beauv.
英名 Little Lovegrass

别名星星草。一年生草本。形态与大画眉相似。只是茎秆更柔弱,圆锥花序更开张,疏松。抗逆性强,适应性广,颖果近矩形,种子落地自繁。幼苗时生长缓慢,拔节后加快。草质柔软,家畜喜食,特别是黄土高原秋季高温多雨时,因生长迅速而有利于季节性放牧。

An annual grass, reproduced from seeds. It is commonly found in farmlands, worn-lands and roadsides, damaging dry-land crops. It is distributed over all parts of the country. The feeding value is high. The texture is soft, the palatability is good with sheep and goats. Horses will graze after frost.

385-2 小画眉草花序　　385-1 小画眉草

386 苇状看麦娘

学名 *Alopecurus arundinaceus* Poir
英名 Tall Alopecurus

别名大看麦娘。禾本科多年生草本。具根状茎。秆直立,多单生少丛生。圆锥花序圆柱状,灰绿色,成熟后为黑色。颖果纺锤形。种子萌发快,分蘖旺盛,适应性广,耐旱性强。在我国"三北"地区分布较多。幼嫩期家畜喜食,抽穗开花期牛、马等大家畜尤为喜食。内蒙有人工栽培种,播种当年生长缓慢,第二年开始利用,刈割晒制的青干草是家畜越冬的优质饲草。

An perennial herb reproduced from seeds. It is commonly found in farmlands and field sides, a weed of wheat, rapeseed and some others crops. It is distributed in south and central Loess Plateau as well as southern China, such as Hubei, Jiangsu, Zhejiang, Guangdong. The palatability is good with horses, cattle, sheep, etc.

386 苇状看麦娘

387　狼尾草

学名　*Pennisetum* alopecuroides (L.) Spreng
英名　Pennisetum

多年生草本。具根状茎。茎秆单生或丛生，直立或斜升。穗状圆锥花序呈圆柱形，直立或弯曲主轴无毛或有微毛。叶鞘较松弛，叶舌有长纤毛，叶片条形。小穗单生于由刚毛状小枝组成的总苞内，成熟时一起脱落。狼尾草是一种广旱生植物，东起辽河平原西至河西走廊。以种子和根茎繁殖，分蘖力很强，再生性良好，是一种优良的牧草，茎叶多而柔软，可食性高，反刍家畜喜食。

A perennial grass, it is spread by seeds and rhizome. It is commnly found in field ridges, roadsides, orchards and nursery gardens, distributed over various parts of the Loess Plateau except some northwestern regions. It is a good grass before flowering, the most domestic animals like to graze it.

387-1　狼尾草

387-2　狼尾草花序

388　金色狗尾草

学名　*Setaria* glauca(L.) Beauv.
英名　Golden Bristlegrass

禾本科一年生草本。茎秆直立，有时基部倾斜，节部生根。下部叶鞘扁平具脊，上部圆形，光滑无毛。叶舌为一圈，有柔毛。叶片下面光滑，上面粗糙。穗状圆锥花序，通常直立。穗上刚毛密集，呈金黄色。小穗椭圆形，先端尖锐。种子与外稃等长，黄灰色。适宜于温暖环境和微碱性土壤，分布广泛。属上等草，叶量大，本身含水高，家畜喜食，而失水也快，便于调制青干草。

An annual grass reproduced from seeds. It is commonly found in rather moist farmlands, creek and roadsides, weed in dry-land crops. It is distributed throughout the country. It is a good forage and palatable for most domestic animals, especially before flowering.

388-2　金色狗尾草花序

388-1　金色狗尾草

389 狗尾草

学名 *Setaria viridis* (L.) Beauv.
英名 Green Bristlegrass, Green Foxtail

一年生草本。疏丛型牧草,茎秆直立或基部膝曲上升。叶片条状披针形;叶鞘光滑,鞘口有柔毛;叶舌具长纤毛。圆锥花序紧密呈圆柱状,直立或微弯曲。刚毛外露,长而硬,绿色或紫色。小穗椭圆形,2或数枚簇生,成熟后与刚毛分离且先于刚毛脱落。第一颖卵形,长度不及小穗的一半;第二颖与小穗等长。第一外稃空虚与小穗等长;第二外稃革质内包种子。种子常椭圆形,先端钝,具细点状皱纹。是一种优良牧草,无论干、鲜,家畜均喜食。

389-1 狗尾草

389-2 狗尾草花序

An annual grass reproduced from seeds. It is commonly found in farmlands, wastelands and roadsides, and distributed throughout the country. The feeding value is high. Cattle like to graze it year around, and the other animals can graze this plant too.

390 大狗尾草

学名 *Setaria faberi* Herrm
英名 Faber Bristlegrass

一年生草本,种子繁殖。茎秆疏丛生,坚硬而高大,直立或基部膝曲,有支持根,高50~120厘米。叶鞘松弛,无毛;叶舌膜质,具短纤毛;叶片无毛或叶面具疣毛。圆锥花序圆柱状,稍弯垂;小穗椭圆形,先端尖,含1~2花;刚毛粗糙。种子椭圆形,先端尖,具横皱纹,成熟后背部极膨胀隆起。主要生于农田、地埂或荒地,分布在黄土高原旱作农业区,草质优良,无论干、鲜,马、牛、骡、驴最为喜食。

390-1 大狗尾草

390-2 大狗尾草花序

An annual grass reproduced from seeds. It occurs in farmlands, wastelands and hill-slopes. It is distributed from northeast China to the Yangtze River valley. It is a very good forage for horses, mule, donkey and cattle.

391 谷莠子

学名 *Setaria viridis* (L.) Beanv.
英名 Major Bristlegrass

一年生草本。与狗尾草相似,但明显比狗尾草植株高大,叶宽、茎粗,穗长、穗粗、穗弯曲度大,小穗长,松散。种子长圆形,比狗尾草种子大而饱满,种子淡黄色不及狗尾草种子色深。牧草品质优良,家畜喜食,尤其是骡、马等大家畜。

An annual grass reproduced from seeds. It is commonly found in farmlands and wastelands, distributed throughout the Loess Plateau. The texture of the grass is soft, with a large leaves, palatability is good and the animals are readily to graze all year around.

391-1 谷莠子　　　391-2 谷莠子花序

392 荻

学名 *Miscanthus sacchariflorus* Hook
英名 Amu Silvergrass

别名苫房草、红柴、红刚芦、芒草、红毛公。多年生草本。根状茎粗壮,横走。秆直立,叶片宽条形,边缘有小刺,叶鞘光滑,叶舌长。圆锥花序顶生,扇形。小穗成对生于各节上,一柄长,一柄短,含2小花,仅第二小花结实。荻为高大的多年生禾草。常野生于干山坡、撂荒地、河漫滩、固定沙丘群,耐瘠薄,繁殖力强。由于茎叶粗硬,适口性较差,开春幼嫩,牛、羊可食,秋季在一般情况下仅食其叶。

A perennial grass that propagated mainly by rhizome. It is commonly found in hill-slopes, wastelands, telands, river beaches and around sand dunes. It is distributed in north, northeast and northwest of China. Cattle and sheep like to graze it in early spring, and it can be made as hay before flowering.

392-1 荻　　　392-2 荻花序

393 芦苇

学名　*Phragmites communis*
英名　Common Reed

别名芦、苇、葭。多年生草本。根状茎粗壮发达，喜水、耐碱，株高 30 厘米到 3~4 米，生长的高矮与水分密切相关，在内陆盐碱地和干旱处植株矮小，外部形态差异甚大。叶鞘无毛或被细毛，叶舌短，叶片扁平，边缘粗糙，披针形，两列。夏季开花，圆锥花序，小穗含 4~7 朵小花。分布于我国温带地区，生长在河岸、道路旁、弃耕地、池沼或盐结皮地。一种良好的保土固堤防风植物，也可盖房、编席。作为饲草，幼嫩时可放牧，成熟后可制成过冬的青干草，家畜喜食。

A perennial grass that propagated mainly by rhizome. It is commonly found nearby villages, gullies and wet lands, and distributed over Loess Plateau as well as around the Yellow River and Yangtze River valleys. Cattle and sheep like to graze leaves as fresh or dry.

393-1　芦苇　　　　　393-2　芦苇花序

394 假苇拂子茅

学名　*Calamagrostis pseupophragmites*
英名　False-reed Reedgrass

根茎型多年生草本。茎秆粗壮，直立或斜升，分枝簇生，光滑无毛。叶片条形，扁平或内卷，边缘及上面较粗糙。圆锥花序呈长圆状披针形或长纺锤形，较开展，成熟后看似蓬松。小穗狭披针形，第二颖略短于第一颖，外稃顶端具长芒，基盘有长柔毛，毛长超过小穗。饲用性较差，幼嫩至抽穗期间家畜喜食，开花后迅速变得粗老，适口性下降。

This is a perennial grass, propagated mainly by rhizome. It is commonly found in hill-slope grasslands or lower lying damp river sides, and distributed in provinces of north, northwest and northeast China as well as Sichuan and Yunnan province. The feeding value is moderate. Cattle is grazing it before flowering.

394-1　假苇拂子茅　　　394-2　假苇拂子茅花序

395　长芒草

学名　*Stipa bungeana* Trin.
英名　Bunge Needlegrass

别名本氏针茅。多年生草本，须根发达坚韧，外具沙套。茎秆密丛生，基部膝曲，下部叶鞘中常常有未抽出的小穗，叶片多纵卷成筒。圆锥花序，下部往往包被于叶鞘之中，完全成熟后才伸出鞘外。花序分枝细弱，直立或斜升，上部疏生小穗。小穗含1花，灰绿色或淡紫色。颖边缘膜质，先端呈细短芒。内外稃等长，外稃背面具呈纵行的短毛，顶端具长芒，该芒丝发状，长3~5厘米，有明显的关节，2回膝曲而扭转。黄土高原北部干旱草原的标志性牧草，抽穗前家畜喜食。

A perennial grass reproduced from seeds. It is commonly found commonly in dry hill-slopes, wastelands, field sides and roadsides. It is distributed over Loess Plateau, especially in north and northwest. The feeding value is quite high with livestock grazing in all seasons. Caryopsis with callus can be harmful to sheep.

395-1　长芒草　　395-2　长芒草花序

396　黄背草

学名　*Themeda japonica* (willd.) Tanaka.
英名　Japanese Themeda

别名菅草、红山草。多年生草本。须根粗壮。茎秆直立，高1米左右。叶鞘紧密裹茎，通常被硬疣毛。叶片条形，有叶舌。圆锥花序较狭窄，长30~40厘米；总状花序长15厘米，下托有无毛的佛焰苞状总苞。每一总状花序有7枚小穗，下面两对多不孕。生于干旱山坡向阳面，在西北海拔500~1000米的地方能成为群落的优势种，南方云贵高原也有分布。在天然草地春夏之交家畜乐于采食。种子基盘有坚硬的锥刺，可刺入家畜皮下或口腔，引发炎症。

A perennial grass reproduced from seeds. It is a common plant found in dry hill-slopes, field sides and gardens. It is distributed over various regions except Xinjiang, Qinghai, Tibet and Inner Mongolia. Its palatablilty is high when it is tender and green, after it is dry.

396-1　黄背草

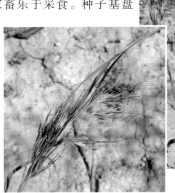

396-2　黄背草花序

397 苏丹草

学名 *Sorghum sudanense* (Piper.) Stapf.
英名 Sudangrass

别名野高粱。一年生草本植物。须根，根系发达。茎直立，呈圆柱状。分蘖力强，侧枝多，一株一般有15～25个，最多50～100个。叶宽线形，色深绿，表面光滑。叶鞘稍长，全包茎，无叶耳。圆锥花序，较松散，分枝细长，每节着生两枚小穗，一无柄，为两性花，能结实。一有柄，为雄性花，不结实。颖果扁卵形。籽粒全被内外稃包被，种子颜色依品种不同而有黄、紫、黑之分。苏丹草原产于非洲的苏丹高原，解放前引进我国，现全国各地均有较大面积的栽培。其蛋白质含量居一年生禾本科牧草之首，多种家畜喜食，也是养鱼的好饲料。

397 苏丹草

An erect annual grass reproduced from seeds.It is suitable for seeding on sands and clay soil, can resist cold, drought, and alkali soil conditions.It is distributed over northern and southern China.It is readily grazed by domestic animals.It can make silage for lactating animals in winter.

398 先锋高丹草

先锋高丹草是饲用高粱与苏丹草杂交育成的一年生禾本科牧草。茎秆直立，高1.2～2米，分蘖能力强。叶片宽长且多。具有营养生长期长，可刈割次数多，产量高；粗蛋白营养丰富，消化率高等特点。其粗蛋白质含量比苏丹草高40%；具有的褐色中脉特性，使其营养物质消化率大为提高，比苏丹草高15%。适口性好，可用于喂养各种畜禽和草食性鱼类。耐旱，对土壤要求不严。在庆阳站生长约170天左右，每667米^{2}产青贮鲜草高达8吨。适口性好，越冬期间牛、羊喜食。

It is a hybrid plant from sorghum and Sudan-grass.It grows widely in farmlands throughout the Loess Plateau and north of China.The yield and feeding value are high.It can be made as silage for cattle (beef and dairy) and sheep in winter.

398-1 先锋高丹草　　398-2 先锋高丹草花序

399　甜高粱

学名　*Sorghum saccharatum*
英名　Sugar sorghum

别名杂交（甜）高粱、糖高粱。一年生草本。高度在2~3米之间，根系非常发达，分蘖多，抗旱能力较玉米强，刈后可再生。茎秆含糖量高而多汁，叶片大小与玉米类似，但茎秆较细。在庆阳黄土高原试验站旱地种植鲜草产量为每666.7米^2 5~6吨，是黄土高原小型养畜场和农户理想的青贮作物。青贮后是解决奶牛、肉牛、绵羊等家畜冬春季节的优质饲草。

An annual cultivated crop distributed in northern China.The yield is very high.The nutritive value and palatability is increased when make into silage.Cattle (beef and dairy) and sheep will graze it in winter.

399-1　甜高粱　　　　399-2　甜高粱分蘖

五十三　莎草科

400　白颖苔草

学名　*Carex Krigescens*
英名　White~scaly Sedge

多年生小草本。具细长的匍匐根茎，高10~15厘米。秆三棱形，基部具纤维状的叶鞘。叶线形，无毛，质较硬。花序穗状，单生，卵圆形，由数个至数十个小穗组成。小穗雄雌顺序，卵形至狭卵形，鳞片卵形，先端钝，锈色，具白色膜质边缘，背面中肋绿色。囊包卵圆形，先端渐狭为短喙。小坚果卵形。生于海拔300~1 300米的路旁、田边、山麓、荒地及干旱山坡，有时也入侵农田和果园。良好的放牧草，草食家畜均喜食。

A perennial small herb,spread by seed and by rhizome.It occurs in field sides,roadsides and farmlands,distributed in north and northwest China.The feeding value is high.Sheep,goats and cattle like to graze all year around.

400-2　白颖苔草花序　　　400-1　白颖苔草

五十三 莎草科

401 香附子

学名　*Cyperus rotundus* L.
英名　Nutgrass Flatsedge

多年生草本。有较长的匍匐根状茎和块根。茎秆散生，直立，呈锐三棱形。叶基生，短于茎秆，叶鞘基部略带棕色。叶状苞片 3~5，下部的两三片长于花序。聚伞花序单生于长侧枝，上面着生多条长短不齐的辐射枝，每枝有若干小穗，小穗排列成伞状。小穗轴具白色透明的翅，小穗细条形，具多花。鳞片卵形或宽卵形，背面中间绿色，两侧紫红色。小坚果三棱状长圆形，暗褐色，具细点。根茎和种子均可繁殖。良好的放牧草，家畜喜食。

A perennial herb reproduced by tuber and by seeds.It grows in damp farmlands and wastelands, and is vicious weeds in paddy fields.It is distributed in southern Loess Plateau as well as southern China such as Guangdong and Yunnan.The feeding value is moderate for cattle and water buffalo.

401-2　香附子花序　　　401-1　香附子

402 藨草

学名　*Scirpus triqueter* L.
英名　Common Bulrush

多年生水生植物。散生，根状茎细，匍匐，秆三棱状，平滑。叶鞘膜质，2~3 个，仅最上部的一枚顶端有 1 叶片，苞片 1 枚，三棱形，稍长于花序；长侧枝聚伞花序假侧生，辐射枝有或无，小穗簇生，花序呈头状；小穗卵形，有多数花；鳞片矩圆形或广倒卵形。雄蕊 3，柱头 2。小坚果倒卵形，褐色，平凸状，平滑光亮。生长在河边、沼泽或浅水中。返青早是放牧家畜的优良牧草。

A perennial herb, spread by rhizome and by seed.It occurs in lake sides and shallow water, being a common weed found in rice field. It is distributed in provinces of north and northeast China as well as Shaanxi, Gansu, etc. Thepalatability is poor, cattle would graze.It can be fed for pigs and some fish when very fresh.

402-1　藨草

402-2　藨草花序

403 水莎草

学名 *Juncellus serotinus* (Rottb) C.B.Clarke.
英名 late juncellus

多年生草本。具有长匍匐根状茎。秆散生,直立。叶鞘腹面棕色;叶状苞片3~4,长于花序。长侧枝聚伞花序复出,具4~7条长短不等的辐射枝,每枝有1~3个穗状花序,每个穗状花序有4~18小穗;小穗条状披针形,穗轴有白色透明的翅;鳞片两列,先端钝,背中肋绿色,两侧红褐色。小坚果倒卵圆,平凸状,有突起的细点。生于浅水或湿地,对稻田危害较重。属于中等牧草,家畜采食。

A perennial herb, spread by rhizome and by seed. It occurs in paddy fields or wet lands of water sides. It is distributed in provinces of south and northeast China as well as Shaanxi, Gansu. The feeding value is moderate and various animals like to graze it.

403-1 水莎草   403-2 水莎草花序

404 旋鳞莎草

学名 *Cyperus michelianus* L.
英名 Revolving Glume Flatsedge

别名白莎草。一年生草本。草层低矮,茎秆丛生,茎三棱状。叶基生,叶鞘基部紫红色。苞片叶状3~6枚,长于花序。聚伞花序紧缩成头状,具多数小穗。小穗长圆状披针形,具多数花。鳞片螺旋状排列,先端有短尖,背部具绿色龙骨状突起,边缘黄白色。坚果小,长圆形,有三棱,暗褐色。零星分布于沟壑和流水两旁的湿地,路边及低洼农田也有。属于放牧型牧草,草质良好,家畜喜食。

An annual sedge, reproduced from seeds. It grows in wet lands, forest edges and shallow water. It is distributed in gullies or damp of Loess Plateau. The feeding value is moderate and cattle, horses and sheep are willing to graze.

404-1 旋鳞莎草   404-2 旋鳞莎草花序

405　两歧飘拂草

学名　*Fimbristylis dichotima* L.
英名　Dichotomous Fimbristylis

一年生草本。茎秆直立，丛生。叶片狭条形，略短于秆。叶鞘淡棕色，鞘口戟形。苞片3～4，其中有的长于花序。长侧枝聚伞花序复出，小穗单生于枝顶，长圆形或近卵形，含多数花。鳞片卵形，褐色，有色泽，先端有短尖。小坚果宽倒卵形，淡黄色，有褐色短柄。多生于沼泽、草甸或潮湿的地方，放牧家畜较喜食。

An annual herb reproduced from seeds.It is commonly found in paddy～fields or damp grasslands.It is a common weed found in various region in the country except Mongolia and Tibet,etc.The feeding value is moderate and horses,cattle,sheep and pigs will graze it when fresh.

405　两歧飘拂草

五十四　天南星科

406　石菖蒲

学名　*Acorus gramineus*
英名　Grassleaf Sweelflag

别名山菖蒲、水剑草。多年生草本，根茎和种子繁殖。全株具香气。硬质的根状茎横走，多分枝。叶剑状条形，两列密生于短茎上，全缘，先端渐尖，有光泽，中脉不明显。4～5月开花，花茎叶状，扁三棱形，肉穗花序，花小而密生，花绿色，浆果肉质，倒卵形。喜阴湿环境，在密度较大的林下也能生长，主要分布于长江流域，黄土高原南缘与秦岭毗邻地带和西藏也有。幼嫩和秋季经霜后牛、羊采食。

A perennial herb, spread by rhizome and by seed.It grows in wetlands, distributed in north and southern China.The palatability is poor, cattle graze it a little.It can be used as a decorative plant.

406　石菖蒲

五十五 忍冬科

407 羽裂莛子藨

学名 *Triosetum pinatifidum* Maxim.
英名 Fleatherycleft Horsegentian

多年生缠绕灌木。茎蔓生、细长、中空、多分枝。嫩株绿色，老株暗红色，叶片大，单叶对生，无柄，卵圆形或长卵形，有 2～3 对称的深裂，裂片上部渐尖，具小尖头。花成对腋生，花初开白色，2～3 天后变黄色，筒状花，有香气，浆果球形，熟时黑色带有光泽。青藏高原和黄土高原均有少量分布。

407-1 羽裂莛子藨果穗

407-2 羽裂莛子藨

A shrub, commonly found in mountain grasslands, alpine meadows and among hassocks along forest fringes. It is distributed in the southwestern part of Loess plateau and Qinghai-Xizang plateau.

408 金银花

学名 *Lonicera japonica* Thunb.
英名 Flos lonicerae

别名金银藤、忍冬。多年生藤本。枝条圆柱形，中空，多分枝，匍匐或半匍匐，老枝表皮黄褐色。单叶对生，叶柄长，密被柔毛，纸质；叶片卵形、阔卵形或卵状披针形，先端短尖、渐尖，基部圆形或近心形，全缘。花成对腋生，花梗密被短柔毛；花冠狭长呈细漏斗状。花色初开白色，继而呈黄色。分布在黄土高原南缘与秦岭接壤的低山丘陵地带。牛、羊喜食其幼嫩的枝叶和花序。

A liana, commonly found in among shrubs, forest fringes or in the cracks between rocks around gullies. It is distributed throughout the country. Cattle, sheep and goats will graze tender branches, leaves and inflorescences.

408-2 金银花花序　　408-1 金银花

五十六 百合科

409 大苞萱草

学名　*Hemerocallis middendorfii* Trautv. et Mey.
英名　Daylily

别名黄花菜、紫萱、忘忧草。根肉质簇生,根端膨大成纺锤形。叶基生,狭长带状,下端重叠,向上渐平展,全缘,中脉于叶下面凸出。花茎自叶腋抽出,圆柱形,茎顶分枝开花,有花 2~10 朵,苞片宽卵形,花被金黄或橙黄色,漏斗形,花被 6 裂。蒴果,革质,椭圆形。种子黑色光亮,花期夏季。生于山坡、草地或栽培。

A perennial herb, spread by seeds and by buds. It grows on arid, hillsides. The feeding value is moderate, sheep and goats like to graze the leaves.

409　大苞萱草

410 卷丹

学名　*Lilium lancifolium* Thunb
英名　Lanceleaf Lily

多年生草本。具须根。基部有鳞茎,簇生成圆球状。茎直立,平滑无毛,灰白中略带褐色。叶多数散生于茎,先端渐尖,基部宽而无柄,全缘。花单一或数朵,生于茎顶,开花时花冠 6 片反卷,淡红黄色,上面具多数不规则的深色斑,有芳香气味。蒴果,近圆形,黑色。幼嫩时放牧家畜尚可采食。

A perennial herb that propagates by bulbs or by seeds. It grows in mountain me-adows, mountain forest edges and field ridges in the Loess Plateau. The feeding value is low and no livestock like it except in winter.

410-1　卷丹　　　**410-2　卷丹花序**

411 天 蒜

学名　*Allium paepalanthoides*
英名　Longbeak Onion

别名小蒜。多年生草本。具球形鳞茎,外皮纸质,白色。有时,鳞茎周围分生 1~5 个球形或半球形小鳞茎,小鳞茎可长成新的植株,鳞茎的下端有发达的须根。叶片半圆形,3~6 枚,内侧常具一明显的纵沟,绿色,通常上覆白粉,手触之则粉脱落。花葶单生,中空,直立,高 10~25 厘米,由叶鞘包被,叶鞘膜质。伞形花序生于葶顶,近球形或半球形,总苞白色,膜质,小花梗等长,花淡红紫色。蒴果,近球形,有棱。全国各地均有分布,主要生于海拔 400~1 600 米的山坡草地、田埂。在阴湿而肥沃的果园生长茂密,对家畜有开胃作用,放牧家畜均喜食,但不能连续采食,是马、牛、羊的抓膘草。

A perennial herb, it is propagated by bulbs. It is commonly found in orchards, nursery gardens, wastelands and farmlands, distributed around the Yangtze River valley and north part of China. It can be used as vegetables, sheep and goats like to graze it.

411-1　天蒜

411-2　天蒜花序

412 碱 韭

学名　*Allium mongolicum* Regel
英名　Mongolian Onion

别名蒙古葱、蒙葱、沙葱。多年生草本。数个鳞茎簇生在一起,鳞茎细长,圆柱形,外皮纤维质,黄褐色。花葶圆柱形。叶基生,圆柱形或半圆柱形,肉质,有灰绿色薄粉层,手触摸时变成绿色。伞形花序球形或半球形,多花,淡玫瑰色。分布于蒙古南部荒漠戈壁地带,典型的旱生植物。是一种季节性放牧草,家畜主要在果实前利用,因具辛辣味,不能长时间单一采食。混合采食可增加食欲,防止某些疾病。羊、驼喜食,食后无虑;牛不太喜食;马多食后会上火。

It is a xerophyte perennial herb, grows on deserts steppes, sands and arid hillsides, distributed in north and northwest China. The leaves and flowers are edible. It is palatable for most livestock and the feeding value is high.

412-2　碱韭花序

412-1　碱韭

413　秦岭野韭

学名　*Allium tuberosum* Rottl
英名　Branchy Onion

多年生草本。具倾斜的横生根茎。鳞茎狭圆锥状，簇生，外皮淡黄褐色，网状纤维质。叶基生，狭条形，扁平，边缘平滑。花葶圆柱状，常具棱，宿存。花序伞形，花梗长于花被，花白色或带红色。蒴果倒圆锥状球形，具棱。在"三北"各地的山坡、平原、农田周围、天然草地均有，蛋白质、必需氨基酸、无氮浸出物、微量元素等含量高。家畜均喜食，常见放牧畜首先采食野韭菜，然后采食其他牧草。

A perennial herb that grows on gravel sloping fields and wastelands.It is distributed in provinces of north, northwest and northeast China as well as Mongolia.It can be used for vegetables, flowers and spice.Sheep, goats, horses and cattle like to graze this excellent forage.

413　秦岭野韭

414　野　韭

学名　*Allium macranthum*
英名　Largeflower Onion

别名山韭。多年生草本。鳞茎不膨大，外被白色膜质叶鞘。须根长，略膨大，肉质。叶基生，3～5枚，线形，基部抱茎。花葶单一，直立，伞形花序顶生，花紫红色，钟状。蒴果。主要分布在我国的西北、西南以及西藏等地，生于海拔2 000米以上的高山草丛。马、牛、羊乐食，是一种调胃、健胃、增进食草量的理想牧草之一，虽然数量少，但是在草原上有着特殊意义。

A perennial herb, spread by seeds and by crown buds.It grows on sands and arid hillsides of deserts, distributed over northern China.It can be used as vegetables, flowers and spice as well as excellent forage. Sheep, goats and cattle like to graze it as will horses..

414　野韭

415 天门冬

学名 *Asparagus sprengeri*
英名 Iran Asparagus

别名天冬草、武竹。多年生草本，种子繁殖，也可分根繁殖。具肉质根，纺锤形。茎丛生，细软，多分枝，蔓性下垂，叶退化成鳞片状，有棱或狭翅。叶状枝3枚一簇，三棱形，镰刀状，基部具小刺。花腋生，花序总状，小花淡绿色至白色，略带淡红色晕斑，1～3朵丛生。浆果球形，成熟后鲜红色，具1粒种子。耐旱，也较耐阴。茎叶可作饲料，全株柔软，属于中嗜牧草，放牧家畜喜食。

A perennial herb, spread by seeds and by crown buds. It exists under forests and hassocks in the mountain valleys. It is distributed in south of Loess plateau. Animals would graze its branches and leaves as fresh or dry.

415-1 天门冬　　　415-2 天门冬果

416 戈壁天门冬

学名 *Asparagus gobicus* Ivan.ex Grubov
英名 Desertliving Asparagus

半灌木。根稍肉质。茎上部通常迥折状，中部具纵向剥离的白色薄膜；分枝亦迥折状，略具纵凸纹，疏生软骨质齿。叶状枝每3～8枚成簇，通常下倾或平展，与分枝成钝角，近圆柱形，略具数棱叶鳞片状，基部具短距，无硬刺。花每1～2朵腋生，单性，雌雄异株，关节位于近中部；雄花：花被片6；花丝中部以下贴生于花被片上；花药矩圆形；雌花略小于雄花。浆果球形，成熟时红色。

It is a semi-shrub, and grows on sands and dry gravel riverbeds in deserts. It is one of characteristic species of desert steppes in northwestern Loess Plateau. The feeding value is moderate, sheep and goats will graze young plants.

416-2 戈壁天门冬果　　　416-1 戈壁天门冬

417 芦 笋

学名 *Asparagus officinalis* L.
英名 Common Asparagus

别名石刁柏、龙须菜。多年生草本。根系发达，须根粗壮，根颈处分蘖旺盛，初出茎鲜嫩如笋，故亦称芦笋。茎秆直立，中空，光滑无毛，多分枝，密集。分枝为上下小，中间大，整个株型如纺锤状。叶状枝 3~6 簇生，稍呈针刺状，鳞片状叶基部具短柄。花小，腋生，淡绿色。浆果圆球形，绿色，成熟后变为红色，内含种子 3~5 枚，种子黑色。枝叶家畜喜食。

A perennial herb, reproduced by seeds and by crown buds. It is common found in watersides wetlands or under forest as cultivated or volunteer plant, distributed throughout the country. The feeding value is high. Cattle and sheep like to graze any time

417-1 芦笋　　417-2 芦笋果

418 百 合

学名 *Lilium brownii*
英名 Lily

多年生草本。株高 40~60 厘米。茎直立，不分枝，草绿色，茎秆基部带红色或紫褐色斑点。地下具鳞茎，阔卵形或披针形，白色或淡黄色，多数须根生于球基部。单叶互生，狭线形，无叶柄，叶脉平行。花着生于茎秆顶端，呈总状花序，簇生或单生，花冠较大，花筒较长，呈漏斗形喇叭状，6 裂无萼片，因茎秆纤细，花朵大，开放时常下垂或平伸；花色，因品种不同而色彩多样，多为黄色、白色、粉红、橙红，有的具紫色或黑色斑点。蒴果椭圆形。主要分布于以兰州为中心的黄土高原西北部，以栽培种常见，嫩茎和叶子是羊的饲草。

A perennial herb, it is spreading by bulbs or by seeds, usually cultivated in northwest regions of the Loess plateau. The leaves and tender branches are good forage, sheep and goats will like it.

418-1 百合　　418-2 百合花序

419　山丹花

学名　*Lilium pumilum*
英名　Low Lily

别名山丹百合，山丹丹，山丹子，细叶百合。多年生草本。鳞茎圆锥形或长卵形。茎上生叶，细长纤弱，狭披针形或线形长如松叶。花顶生，通常1~3朵，下垂。花梗长；花被片6，中间凹；反卷，鲜红色或紫红色，披针形，无斑点或有少数斑点；花药5，具红色花粉。蒴果近球形。花期6~8月，果期8~9月。生于向阳山坡；或有栽培。分布于黑龙江、吉林、辽宁、河北、河南、山东、山西。

A perennial herb, it is spreading by bulbs or by seeds, it usually exists in hill, among thickets and forest edge. It is distributed in northern Loess plateau. The feeding value is moderate, cattle, sheep and goats will graze it.

419-1　山丹花　　419-2　山丹花花序　　419-3　山丹花茎

五十七　鸢尾科

420　马　蔺

学名　*Iris lactea* Pall
英名　Chinese Iris

别名马莲。多年生草本。根状茎短粗，须根棕褐色，植株基部具红褐色纤维状的叶鞘残留物。叶基生，甚坚韧，条形。花葶自基部生出，有花1~3朵，蓝紫色。蒴果椭圆形，有6条纵肋，先端有喙。种子多数，近球形而有棱角，棕褐色。全国各地均有分布。初萌芽和经霜后，家畜少食。

It is perennial grass and spread by seeds and by crown. It grows on beach land, saline flood lands and roadsides. It is distributed over northern and southern China. The feeding value is moderate and domestic animals graze after frost in the autumn.

420-2　马蔺花序　　420-1　马蔺

421 射干鸢尾

学名 *Gladiolus* L.
英名 Blachberrylily Rhizome

唐菖蒲属，分布于地中海区和南非洲，我国引入有数种，各大城市有栽培，剪其花茎供插瓶用，花的颜色种种。球茎有膜被；叶剑状，有时线形或圆柱形；花茎直立，常单生，多少具叶，有大而草质的佛焰苞，每一苞内着生无柄花1朵；花大，左右对称，通常美丽，白色、黄色、紫色、红色或他色；花被管漏斗状，大部分种类向上弯曲；裂片6，上面3裂片较大；雄蕊3，通常多偏于一侧；子房下位，3室，有胚珠多颗，蒴果室裂；

421-2 射干鸢尾花序

421-1 射干鸢尾

种子扁平或有翅。主要分布在西北黄土高原干旱地区，其干燥后的叶片可以加工为家畜越冬的饲草。

A perennial herb, spread by rhizome and by seed. It occurs in marshlands, stream and paddy field sides, distributed over all parts of the country. The palatability is poor, livestock graze a little.

422 膜苞鸢尾

学名 *Iris scariosa* willd.
英名 Scarious Swordflag

别名紫蝴蝶、蓝蝴蝶、乌鸢、扁竹花。多年生宿根性直立草本。根状茎匍匐多节，粗而节间短，浅黄色。基部有数枚短的鞘状叶，边缘膜质，对折，呈镰形向外弯曲，披针形，互相包叠。总状花序1~2枝，每枝有花2~3朵；花蝶形，花冠蓝紫色或紫白色，外3枚较大，圆形下垂；内3枚较小，倒圆形；外列花被有深紫斑点，中央面有一行鸡冠状白色带紫纹突起，雄蕊3枚，与外轮花被对生；花柱3歧，扁平如花瓣状，覆盖着雄蕊。蒴果长椭圆形，有6棱。主要分布北方干旱地区，秋季经霜后家畜采食。

A perennial herb, spread by rhizome and by seed. It exists in roadsides and wastelands, distributed in arid regions of north and northwest China. The feeding value is low with domestic animals grazing after a frost in the autumn as well as winter.

422-1 膜苞鸢尾花序   **422-2 膜苞鸢尾**

五十八 兰 科

423 绶 草

学名 *Spiranthes sinensis*
英名 China Ladytress

多年生小草本。有簇生的肉质块根,萝卜状。茎直立,无分枝,下部着生数片披针形叶。顶生穗状花序较长,呈螺旋状扭转,花小,淡红色,钟状,通常偏于花序的一侧,花具有明显的唇瓣。蒴果。多生于河滩或低海拔草地,零星分布。适口性较好,幼嫩或干燥后家畜均喜食。

A perennial herb, with succulent roots, it grows on swamp meadows or forest edge, distributed over China. The feeding value is low, the most animals do not like it, but can be used for snakebites.

423-2 绶草花序 423-1 绶草

五十九 亚麻科

424 宿根亚麻

学名 *Linum stelleroides* Planch
英名 Wild Flax

别名野胡麻。一年生草本。直根系,入土深,侧根多而细长。茎直立或斜升,多下部分枝。叶片披针形或条状披针形,先端锐尖,基部渐狭,全缘。聚伞花序稀疏,花单生于枝顶或茎枝上部叶腋,有花梗。萼片5,卵形或卵状披针形。花瓣5,倒卵形,蓝色或紫蓝色,稀白色。蒴果球形,顶端5裂,通常内含5~10粒种子,种子扁,长椭圆形或长卵形。草质柔软,放牧家畜喜食。

A perennial herb, resistant to drought, widely spread on grasslands regions, mountains and gravel lands. It is distributed over north China as well as Loess Plateau. Livestock animals will graze fresh stems and leaves.

424-1 宿根亚麻 424-2 宿根亚麻花序

六十 荷包牡丹科

425 地丁草

学名 *Corydalis bungeana* Turcz.
英名 Philippine Violet Herb

多年生或越年生草本，种子繁殖。根细而直，少分枝，淡黄棕色。茎细弱，从基部向四周多分枝，有棱，灰绿色，通体无毛。基叶丛生，有长柄，叶片2回羽状全裂，最终裂片线形，先端钝圆或成短突尖，两面具粉粒，灰绿色。总状花序，生于枝的顶端；苞片叶状，羽状深裂，萼片2枚，鳞片状；花瓣4枚，淡紫色，倒卵状长圆形，上片基部有距，背面有翅。蒴果扁椭圆形，灰绿色。种子扁球形，黑色，有光泽。草食家畜喜食。

425-2 地丁草花序

425-1 地丁草

A biennial or perennial herbaceous plant reproduced from seeds. It occurs in farmlands, roadsides, ditch banks, wastelands and tree nurseries. It is distributed in Gansu, Shanxi, Hebei, Liaoning and some other provinces.

六十一 虎耳草科

426 梅花草

学名 *Parnassis palustris* L.
英名 Wideword Parnassia

多年生草本，种子繁殖。基生叶丛生，卵圆形或心形，全缘；茎生叶1，生于茎中部，无柄，形状与基生叶相似。花单生于茎顶，白色，形似梅花；萼片5，长椭圆形；花瓣5，平展，卵圆形，先端广圆头，全缘；雄蕊5，与花瓣互生，退化雄蕊5，丝状分裂成多条，裂片先端有头状腺体；心皮4，合生，子房上位；花柱短，柱头4裂。蒴果卵圆形。种子多数。生于草甸、河边湿地、林下、水沟旁。分布于我国"三北"地区。

A perennial herbaceous plant reproduced from seeds. It is commonly found in meadows, wet lands, water sides and river banks. It is distributed in north, northeast and northwest China as well as Gansu and Shanxi provinces.

426-1 梅花草   426-2 梅花草叶

六十二 川续断科

427 东北川续断

学名 *Dipsacus japonicus*
英名 Dipsacaceae

427-1 东北川续断　　427-2 东北川续断花序

别名黑老鸦头、山萝卜根。多年生草本。根圆锥形,主根明显,根皮黄色或土黄色。茎直立,多分枝具棱和浅沟,棱上有倒钩刺。基生叶长椭圆形,有长柄;茎生叶对生,倒卵状椭圆形。叶片3~5回羽状深裂,中央裂最大,两侧渐小,先端锐尖,基部楔形,下延成狭翅,边缘有粗锯齿,两面被疏白毛,背脉和叶柄有钩刺。顶生花序,刺球状。苞片多数,螺旋状紧密排列。花萼皿状,4浅裂,外被白色。花冠紫红色,漏斗状,基部呈短细筒,内外被毛。瘦果稍外露。少量零星分布。

A perennial herb, reproduced from seeds. It exists in hill slopes and hassocks in the mountain valleys. It is distributed in southeast Loess plateau and Qinling mountain.

六十三 薯蓣科

428 穿龙薯蓣

学名 *Discorea nipponica* Makino.
英名 Throughhill Yam

别名川龙薯蓣、串地龙。多年生缠绕藤本植物。根茎横走,圆柱形,黄褐色。茎左旋细长,近乎无毛。叶互生,叶片卵形或广卵圆形,5~7浅裂,先端渐尖,基部心形。花黄绿色,小,雌雄异株,穗状花序,腋生,雄花序复穗状,雌花序单一,下垂,绿黄色。蒴果卵形或椭圆形,具3翅,成熟后黄褐色。种子上部具长方形膜质翅。分布于东北、华北及甘肃、陕西等省。家畜喜食。

A perennial trailing herb, spread by crown and by seeds. It grows in hill, among thickets and forest edges and cultivated in farmlands also. It is distributed in southeast and central Loess plateau. The feeding value is high, the leaves, stems and roots are good forage and mostly animals like to graze.

428 穿龙薯蓣

六十四　檀香科

429　百蕊草

学名　*Thesium chinense* Turcz.
英名　China Bastardtoadflax

多年生草本。基部分枝，枝细柔，有棱条，斜上。叶互生，线形而尖。具1脉。花小，单生叶腋，花被钟状，绿白色，具1苞片和2小苞片，5裂，偶4裂；雄蕊与裂片同数，着生于裂口内面，并与裂片对生；子房下位，柱头头状。果实球形；花被宿存，网纹明显。生于山坡、路旁或田野，广布南北各省。

A perennial herb, reproduced from seeds. It exists in hill slopes, roadsides and field sides. It is distributed throughout the country. The feeding value is moderate and it is palatable for grazing livestock when it is fresh.

429-1　百蕊草

六十五　商陆科

430　美国商陆

学名　*Phytolacca americana* L.
英名　Pokeberry Root

别名垂序商陆。一年生草本。高1~2米。根肥大，倒圆锥形。茎直立或披散，圆柱形，有时带紫红色。叶椭圆状卵形或卵状披针形，先端急尖。总状花序顶生或侧生，花梗长4~12厘米，花白色，带红晕。

A perennial herb, spread by root tuber or by seeds. It is suited to fertile and wet lands, commonly found in hill slopes, gullies and among hassocks. It is distributed in southeast Loess plateau and Qinling mountain region.

206-1　美国商陆

206-2　美国商陆花序　　206-3　美国商陆浆果

六十六　无患子科

431　文冠果

学名　*Xanthoceras sorbifolia* Bunge.
英名　Shinyleaf Yellowhorn

别名木瓜。多灌木少乔木。树皮灰褐色,小枝紫褐色,光滑或有短柔毛。奇数羽状复叶,互生,小叶9~19片,矩圆形或披针形,无柄,边缘有锐锯齿。总状花序顶生或腋生,长15~25厘米;小花具细长梗,萼片5,花瓣5,白色有紫红色斑纹。果大,球形或卵圆形,3~4室,每室有种子1~8粒;种子球形,黑褐色。野生种主要分布在北方各省区,以黄土高原的陕西、山西、甘肃东南部及内蒙最多,一般生长在山坡或崖边。近年作为油料作物引种栽培获得了良好的结果。其叶子蛋白质(14.4%)和脂肪(9.6%)含量高,是家畜好饲料。

A shrub, spreading by radical bud and by seed.it exists in forest fringes and gulley sides and is distributed over Loess plateau as well as in northern China.The tender branches and leaves are very good forage for sheep and goats.

431　文冠果

六十七　木犀科

432　迎春花

学名　*Jasminum nudiflorum*
英名　Winter Jasmine

别名金梅、金腰带、小黄花,落叶灌木。繁殖以分株、压条、扦插为主。枝条细长,呈拱形下垂生长,长可达5米以上。侧枝健壮,四棱形,绿色。三出复叶对生,长2~3厘米,小叶卵状椭圆形,表面光滑,全缘。花单生于叶腋间,花冠高脚杯状,鲜黄色,顶端6裂,或成复瓣。花于早春2~3月份先叶开放。主要分布在云贵高原、黄土高原及四川、河南等省。生于海拔700~2 500米的山坡灌丛或石缝中,马、牛、羊喜食其嫩枝叶,为中等饲用植物。

A shrub or perennial herb, spread by crown buds.It usually occurs among thickets of mountains or gullies.It is distributed in the Yunnan-Guizhou plateau,loess plateau as well as Henan,Sichuan provinces.The feeding value is moderate.Horses, cattle and sheep will graze its leaves and fresh branches.

432　迎春花

附录一：有毒有害植物

一 豆 科

1 黄毛棘豆

学名 *Oxytropis ochranthd* Turcz. in Bull. Soc. Nat. Mosc.

多年生草本。主根木质化而坚韧。茎极短，多分枝，通体被土黄色长柔毛。羽状复叶；叶轴上面有沟；托叶膜质，宽卵形，先端急尖，中下部与叶柄贴生；小叶 6～9 对，对生或 4 片轮生，卵形、长椭圆形、披针形或线形，先端渐尖或急尖，基部圆形，嫩小叶密被丝状贴伏长柔毛。花多数，排列成密集的圆筒状的总状花序；花葶坚挺，圆柱状；苞片披针形，较萼长；萼坚硬，几革质，管状。荚果膜质，卵形，膨胀成囊状而略扁。分布于黄土高原西南部及内蒙、四川等地。

1-2 黄毛棘豆花序　　1-1 黄毛棘豆

A perennial legume, It is commonly found on alpine grasslands, mountain forest edges and wastelands. It is distributed in Gansu, Shanxi, Qinghai, Xizang and Sichuan. It is moderate forages, cattle and sheep will graze it.

2 苦 参

学名　*Sophora flavescens*
英名　Foxtail-like sophora

别名山槐子、地槐、草本槐、苦豆根。小灌木。幼枝有疏毛，后变为无毛。羽状复叶，小叶多达 25～30 片，披针形或条披针形，稀为椭圆形，先端渐尖，基部圆形，背面密生柔毛。总状花序顶生，花冠淡黄色。荚果长 5～8 厘米，呈不甚明显的串珠状，疏生柔毛，含种子 1～5 粒。遍布黄土高原，有毒，家畜一般不采食。

A perennial herb that grows on slopes, roadsides and riversides. The palatability is poor and the feeding value is low. Animals do nott graze. Sometimes cattle and sheep like to graze its tender branches and leaves.

2-1 苦参　　2-2 苦参花序

二 毛茛科

3 耧斗菜

学名　*Aquilegia Viridiflora Pall.*
英名　Greenflower Columbine

多年生草本,种子繁殖。茎直立或斜升,细弱,多分枝,密生短腺毛。基生叶具长柄,为1~2回3出复叶,小叶菱状倒卵形,3深裂或3浅裂,上面无毛,下面疏生柔毛,茎生叶较小。花下垂,萼片5,紫色,狭卵形;萼片与花瓣同色,下面有矩,雄蕊多数,心皮5,子房密被短腺毛。骨葖果,种子黑色。该草根部含生物碱较多,不宜作饲草。

A perennial herbaceous plant reproduced from seeds.It is commonly found in grassy hill-slopes and mountain valleys,distributed in east and southeast Loess plateau as well as Shandong and Hebei.It can not be used as forage.

3　耧斗菜

4 茴茴蒜

学名　*Ranunculus chinensis* Bge.
英名　China Buttercup

别名山辣椒。多年生草本。茎直立,茎与叶柄被黄糙毛。三出复叶具柄,叶片宽卵形,中央小叶有长柄,3深裂,侧生小叶柄短。花序具疏花,5萼片,外面疏被柔毛,花瓣黄色,多雄蕊,多心皮。聚合果矩圆形。多生于海拔300~1 600米的潮湿阴坡和沟谷溪水处。有毒,含原白头翁素 Protoanemonin($C_5H_4O_2$)能导致家畜呕吐、下痢、便血、瞳孔放大等,故不宜作饲草。

A perennial herb, spread by seeds and by crown buds.It occurs in grasslands, creek and roadsides, distributed in north, northeast and northwest China such as Gansu, Shaanxi,etc.It is a toxic plant for the animals.

4-2　茴茴蒜花序　　4-1　茴茴蒜

二 毛茛科

5 草玉梅

学名 *Anemone dichotoma* L.
英名 Broocklet Windflower

别名小花草玉梅。多年生草本。根状茎发达,茎直立,基生叶3~5片,具长柄,叶片轮廓掌状,基部心形,3全裂,中央裂片宽菱形、卵状菱形或狭倒卵状披针形,上部有不明显的3浅裂,先端急尖,基部长楔形,两侧裂片稍宽,斜倒卵形,不对等2深裂。花单一,腋生,有时为聚伞花序,苞片3,披针形。萼片通常5,白色。聚合果近球形。秦岭山脉多见,黄土高原南缘零星分布。常生于1 500米以上的山坡草丛中,不作饲用。

5-2 草玉梅花序

5-1 草玉梅

A perennial herb, it is spreading by seeds and by radical bud. It grows among hassocks, under forest and wastelands, and distributed in southern Loess plateau. Usually livestock animals do not graze.

6 腺毛唐松草

学名 *Thalictrum foetidum* L.
英名 Glandularhairy Meadowrue

多年生草本,种子繁殖。无毛或幼时有短柔毛。茎直立,上部分枝或不分枝。叶互生;叶柄短,有鞘;托叶膜质,褐色;茎中部叶为3~4回三出近羽状复叶,有短柄;小叶草质,菱状宽倒卵形、卵形或近圆形,3浅裂,裂片全缘或有2~3疏齿,上面脉稍凹陷,疏被腺毛,下面脉稍隆起,沿脉生短柔毛和腺毛。圆锥花序,具少数或多数花;花两性;花梗细,被白色短柔毛和腺毛;萼片4~5,花瓣状,卵形,淡黄绿色,外面常有疏柔毛,花瓣无。瘦果倒卵形,扁平。不宜作饲草。

A perennial herbaceous plant reproduced from seeds. It is commonly found in grassands of mountain areas and forest edges, and distributed in northwest China and Loess plateau. Domestic animals will graze it after a frost in autumn.

6-1 腺毛唐松草   6-2 腺毛唐松草花序

7　翠　雀

学名　*Delphinium grandiflorum* L.
英名　Bouquet Larkspur

别名大花飞燕草。多年生草本。茎直立或斜生，有腺毛。基生叶大而多，具长柄；茎生叶小而少，具短柄或无柄。叶多圆肾形，掌状3全裂，裂片再细裂，小裂片条形，两面均无毛。总状花序，花稀少，大而美观。萼片5，花瓣条状，蓝色或紫蓝色，有距。花瓣2，短于萼片，心皮3，有毛。蓇葖果，被短柔毛，边缘具干膜质齿。全草有毒，不能饲喂家畜。

A perennial herb reproduced from seeds. It grows on meadows, mountain and sandy grasslands, distributed throughout the Loess Plateau. Usually all animals do not graze.

7-1　翠雀　　　　7-2　翠雀花序

8　腺毛翠雀

学名　*Delphinium grandiflorum* L. var. *glandulosum* W. T. Wang.

多年生草本，根芽和种子繁殖。成株高35～65厘米。基生叶和茎下部叶具长柄。叶片圆肾形，3全裂，小裂片线状披针形至线形，两面均有毛，茎生叶较少，叶柄向上渐短，以至无柄。总状花序，花稀疏，花序轴和花梗密被黄色腺毛和反曲的微短柔毛，小苞片条形或钻形。萼片5，花瓣状，蓝色或紫蓝色，有距，距通常较萼片长。花瓣2，短于萼片，有黄白色块斑，雄蕊多数，退化雄蕊2，心皮3，有毛。蓇葖果，被短柔毛，种子扁片状，边缘有干膜质翅。分布于甘肃、青海、陕西、山西、河北、河南、安徽、江苏等地。列入有毒植物，不宜饲喂家畜。

A perennial herb, it is reproduced from seeds, and occurs in hills, gullies and field-sides. It is distributed in Gansu, Shaanxi and Ningxia, etc. It is a toxic plant for the animals. The livestock usually do not graze.

8-1　腺毛翠雀　　　　8-2　腺毛翠雀花序

9 石龙芮

学名 *Ranunculus sceleratus* L.

一年生草本，种子繁殖。茎直立，粗壮，稍肉质，有分枝。基生叶或近基部的叶具长柄；叶片宽卵形，3~4深裂，顶生裂片菱状卵形，全缘或有疏齿；茎生叶互生，中部的叶有柄，3裂，裂片狭长圆形，最上部的叶无柄或近无柄，叶片分裂或不分裂。花序有多数小花组成，萼片5，淡绿色，有毛，花瓣5，鲜黄色。聚合果长圆形，瘦果宽卵形，扁平。分布在1 000米及以下的沟边、溪旁，我国均有零星分布，以南方为多见。该草含有毒的毛茛油（Protoanemonin），马、牛、羊采食后，会引起胃肠炎、下痢、便血等不良症状。

An annual herbaceous plant reproduced from seeds.It occurs in rice fields,vegetable gardens,wet lands and ditch sides.It is distributed throughout the country.It is a poisonous plant and all livestock do not graze.

9-1 石龙芮花序

9-2 石龙芮

三 罂粟科

10 秃疮花

学名 *Dicranostiama leptopodum*

越年生草本。通体含淡黄色汁液，具毛，有毒。基生叶莲座状，多数，具柄，背面有白粉。叶片羽状全裂或深裂，2回裂片边缘疏生小齿，茎生叶小而无柄。聚伞花序，花冠鲜黄色，花萼2片，早落，花瓣4枚。子房1室，蒴果细筒形，种子肾形，黑褐色，具粗网纹。主要分布于黄土高原西南，生长在山坡草地、弃耕地、崖边或路旁，有时入侵农田。

A biennial or perennial herb reproduced from seeds.It is commonly found in grassy hill-slopes and orchards,distributed throughout the Loess Plateau as well as northest China,such as Qinghai and Tibet.The animals can graze this plant,which will cause the color of milk changed.

10-1 秃疮花

10-2 秃疮花花序

11　灰绿黄堇

学名　*Corydalis adunca*
英名　Greyqreen Corydalis

一年生草本，种子繁殖。茎多自基部分枝，中间直立，周边斜升，有纵棱，全枝被灰色粉粒。叶片基生并茎生，有柄，2～3回羽状全裂，总状花序生于枝顶，较长，花多，排列稀疏，小花具柄，花有距。花瓣黄色，萼片较小。蒴果条形，略带串珠状。分布在甘肃南部、陕西及江苏、浙江等地。有毒，不能饲用。

An annual herbaceous plant, commonly found in wastelands of rocky hills, orchards, forests and damp ditch sides. It is distributed in southern Loess plateau as well as Jiangsu, Zhejiang and other provinces. Usually, most animals do not graze it.

11-1　灰绿黄堇　　11-2　灰绿黄堇花序

12　白屈菜

学名　*Chelidonium majus* L.
英名　Celandine

别名山黄连。多年生草本，种子繁殖。茎直立，有分枝，高30～60厘米，有黄色液汁。叶互生，羽状分裂，深裂或全裂，上面无毛，背面疏生短柔毛，有白粉。花数朵，呈伞形花序，萼片2，花瓣4，倒卵形，黄色；雄蕊多数，雌蕊无毛，1室。蒴果条状原筒形。含有毒植物碱（Chlidonin、Protopin 等），食后会引起家畜吐泻和中枢神经麻痹而死。

A perennial herbaceous plant, it occurs in fields, roadsides, hill slopes on grasslands along forest edges in mountain valleys. It is distributed in north and northeast China as well as Gansu, Shaanxi, Xinjiang and Sichuan. It is a poisonous plant for animals.

12-1　白屈菜　　　　　　　　　　12-2　白屈菜花序

四 大戟科

13 泽漆

学名 *Euphorbia helioscopia*
英名 Sunn Euphorbia

根据花序和苞片生长的形状，也叫"五朵云"。越年生或一年生草本。全株含乳汁。茎自基部分枝，直立或斜升，圆柱形，通常无毛。叶互生，近无柄，叶片倒卵形或匙形，叶缘中部以上有细锯齿，下部全缘。茎顶部有5片轮生的叶状苞，较叶稍大。总花序顶生，伞梗5，每梗再生3个小伞梗，各小伞梗又分二叉。花小，无花被，单性，雌雄同序；总苞先端4浅裂。蒴果无毛，种子卵形，灰褐色，有网状凹陷。除青藏高原外，我国各省都有分布，生于山沟、路旁、荒野以及果园和农田。有毒，无论干、鲜家畜均不能食。

13-2 泽漆花序

13-1 泽漆

A biennial or annual herb reproduced from seeds. It is commonly found in gullies, wilderness and farmlands, distributed in all regions of the country except Tibet. It is a toxic plant to the most animals.

14 甘遂

学名 *Euphorbia lunulata* Bunge
英名 Kansui

别名猫儿眼。多年生草本。茎多分枝，无毛，基部坚硬。叶片狭条形，先端钝圆，两面无毛，全缘，近无柄，密生于枝条周围。花序基部呈扇状半月形至三角状肾形。总状花序顶生，有5~6个伞梗，每伞梗有2~3个分枝。杯状聚伞花序，无毛，顶端4~5裂，裂片间腺体新月形，枝顶端的花序内仅含雄花，两侧的含多数雄花而仅含1雌花。子房3室，3花柱，分离。蒴果扁球形，无毛，种子光滑。有毒，不能作为家畜饲草。

A perennial herb reproduced from seeds. It occurs in hill-slopes, valleys and sunny river banks. It is distributed in provinces of north China as well as Heilongjiang, Liaoning, Shandong and Jiangsu. It is a toxic plant to the most animals.

14-1 甘遂

14-2 甘遂花序

15 大　戟

学名　*Euphorbia pekinensis* Rupr.
英名　Beijing Spurge

多年生草本。根圆锥状；茎直立，被白色短柔毛，上部分枝，叶互生，几无柄，矩圆状披针形至披针形，肉质，全缘。背面稍被白粉。植株各部有丰富的白色乳汁。总花序通常有5伞梗，基部有卵形或卵状披针形苞片5枚轮生；杯状花序总苞坛形，顶端4裂，腺体椭圆形无花瓣状附属物，子房球形，3室，花柱3，顶端2裂。该草有毒，不能饲喂家畜。

A perennial herb, spread from seed or by crown bud. It occurs in hill slopes, roadsides and wastelands, distributed throughout the Loess plateau. It is a toxic plant, usually livestock animals do not graze.

15-1　大戟　　　　**15-2　大戟花序**

五　瑞香科

16 狼　毒

学名　*Stellera chamaejasme* L.
英名　China Stellera

多年生草本。具圆形粗壮的肉质根。茎丛生，直立，无分枝。叶小无柄，互生。叶片椭圆状披针形，全缘，无毛，密生于茎秆。头状花序顶生，花淡红色或白色，花被筒细长。果实椭圆形。遍布黄土高原，主要分布在海拔1 000～1 500米的向阳干燥的山坡，是一种干旱标志性植物。有毒，不能饲用。

It is a perennial herb with succulent roots, exists on sunny hill side in dry area. It is distributed in provinces in northeast and southwest China as well as Gansu, Shaanxi, Ningxia, etc. It is a toxic plant for the animals, livestock do not graze.

16-2　狼毒花序　　　　**16-1　狼毒**

六 茄 科

17 曼陀罗

学名 *Datura stramonium* L.
英名 Jimson weed

一年生草本。茎直立,粗壮,中空,表面光滑无毛或幼株具疏毛,上部枝条呈二叉分枝。叶互生,上部为假对生,具长柄,叶片卵形或宽卵形,顶端尖,基部楔形,全缘或有波状齿。花单生于叶腋或枝条分叉处,花萼筒状,顶端 5 裂。花冠漏斗状,白色或淡黄色,也有紫色种。蒴果直立,硕大,卵圆状,表面有长短不等的硬刺,成熟后 4 瓣裂。种子多粒,扁肾形,黑褐色。属有毒植物,不能饲喂家畜。

An annual herb reproduced from seeds. It is commonly found in wastelands and farmlands. It is distributed over all parts of the country. It is a toxic plant, the domestic animals do not like it.

17-1 曼陀罗　　　　　17-2 曼陀罗果　　　　　17-3 曼陀罗花序

七 天南星科

18 半 夏

学名 *Pinellia ternata*
英名 pinellia tuberifera tenore

别名三叶半夏、地八豆。多年生小草本。高 15~30 厘米。块茎近球形。叶出自块茎顶端,叶柄长 6~23 厘米;一年生的叶为单叶,卵状心形;2~3 年后,变为三出复叶,小叶椭圆形至披针形,先端锐尖,基部楔形,全缘,两面光滑无毛。肉穗花序顶生,花序梗常较叶柄长;佛焰苞绿色;花单性,无花被,雌雄同株;雄花白色,雌花绿色。浆果卵状椭圆形,绿色。生于山坡、农田、溪边阴湿的草丛中或林下。我国大部分地区有分布。家畜不食。

A small perennial herb, it is propagated by bulbs and by seeds, occurs in farmlands, wastelands or river sides. It is distributed in southeast Loess plateau and other parts of the country. Animals do not graze.

18 半夏

八 禾本科

19 醉马草

学名 *Achnatherum inebrians* (Hance) Keng.
英名 Inebriate Jijiqrass

多年生草本,种子繁殖。基部分蘖,株高60~100厘米,茎节下贴生微毛。叶片长条形,质地较硬,卷折。圆锥花序紧缩近穗状;小穗灰绿色,成熟后变为褐铜色或带紫色;芒长约1厘米,中部以下稍扭转。颖果圆柱形,成熟后易落粒。分布于内蒙古、宁夏、甘肃、青海、新疆、四川等省区。全草有毒,马、骡采食鲜草达体重1%,在30~60分钟后即可出现中毒症状,口吐白沫、精神沉郁、食欲减退、头耳下垂、行走摇晃、如酒醉状。

A perennial grass reproduced from seeds.It is commonly found in alpine meadows and mountain grasslands,and is distributed in north,northeast and northwest China as well as QinghaiXizang plateau.Livestock generally do not graze this plant.

19-1 醉马草

19-2 醉马草花序

附录二:中文名索引

一 画

一包针 …………………………… 144
一年生早熟禾 …………………… 182
一年蓬 …………………………… 148

二 画

二色补血草 ……………………… 106
二色棘豆 ………………………… 77
二裂委陵菜 ……………………… 46

三 画

三叶半夏 ………………………… 225
三叶酸 …………………………… 84
三齿萼野豌豆 …………………… 79
土庄 ……………………………… 49
土荆芥 …………………………… 118
大丁草 …………………………… 158
大力士 …………………………… 162
大马蓼 …………………………… 6
大车前 …………………………… 136
大火草 …………………………… 33
大叶龙胆 ………………………… 107
大叶罗布麻 ……………………… 108
大花千里光 ……………………… 164
大花飞燕草 ……………………… 220
大花罗布麻 ……………………… 108
大苞萱草 ………………………… 205
大苞滨藜 ………………………… 19
大画眉草 ………………………… 192
大刺儿菜 ………………………… 146
大狗尾草 ………………………… 195
大看麦娘 ………………………… 193
大籽蒿 …………………………… 170
大麻 ……………………………… 4
大巢菜 …………………………… 60
大戟 ……………………………… 224
大蓟 ……………………………… 146
大蓼 ……………………………… 6
大辣辣 …………………………… 40
大蝎子草 ………………………… 5
山丹子 …………………………… 210
山丹丹 …………………………… 210
山丹百合 ………………………… 210
山丹花 …………………………… 210
山羊豆 …………………………… 62
山米壳 …………………………… 35
山豆花 …………………………… 84
山苦荬 …………………………… 150
山枣子 …………………………… 49
山金饱 …………………………… 95
山油子 …………………………… 118
山茼蒿 …………………………… 143
山茶根 …………………………… 125
山韭 ……………………………… 207
山桃 ……………………………… 51
山黄连 …………………………… 222
山菖蒲 …………………………… 203
山萝卜根 ………………………… 214
山野豌豆 ………………………… 79
山槐子 …………………………… 217
山辣椒 …………………………… 218
山鳘豆 …………………………… 62
千里光 …………………………… 164
千层塔 …………………………… 148
千穗谷 …………………………… 25,27
川升麻 …………………………… 34
川龙薯蓣 ………………………… 214
女青 ……………………………… 109
女娄菜 …………………………… 30
飞廉 ……………………………… 145
小飞蓬 …………………………… 147
小车前 …………………………… 137
小龙胆 …………………………… 106
小叶锦鸡儿 ……………………… 76
小白酒草 ………………………… 147
小白蒿 …………………………… 172

小耳朵草	91
小米草	132
小花草玉梅	219
小花鬼针草	144
小花糖芥	42
小画眉草	193
小刺叉	144
小果亚麻芥	38
小鬼叉	144
小冠花	59
小柴篮子	80
小黄花	216
小巢菜	80
小蒜	206
小蓟	146
小藜	23
马牙豆	62
马苜蓿	58
马齿苋	28
马齿豆	65
马茹	51
马莲	210
马蔺	210
马缨花	81
马蹄针	81
马鞭草	116

四 画

天门冬	208
天冬草	208
天兰冰草	185
天名精	145
天蒜	206
天蓝苜蓿	73
天蔓青	145
天蓼	6
天撤	95
无芒麦	174
无芒雀麦	174
无芒䅟	190
无根草	113
木本委陵菜	52
木瓜	216

木蓝	80
五叶草	59
五加	94
太白岩黄芪	67
太阳花	86
车串串	137
车轮菜	137
车轱辘菜	136
车前	136
车前草	136
戈壁天门冬	208
止血马唐	188
少花米口袋	71
日本菟丝子	113
中亚滨藜	19
中间冰草	185
中间偃麦草	185
中间锦鸡儿	76
中国菟丝子	112
牛毛蒿	155
牛角花	59
牛筋草	189
牛蒡	162
牛膝	25
牛繁缕	29
毛马唐	188
毛花绣线菊	49
毛茛子	61
毛胡枝子	84
毛野豌豆	61
升马唐	188
升麻	34
长叶铁扫帚	71
长芒草	198
长芒野稗	191
长柱金丝桃	98
反枝苋	26
月见草	100
风毛菊	154
风花菜	40
丹参	123
乌茑	211
六月禾	181

文冠果	216
火绒草	159
火媒草	154
孔明菜	143
水红蓼	6
水芙	97
水芹	102
水芹菜	102
水杨梅	48
水剑草	203
水莎草	202
水棘针	118

五　画

甘西鼠尾草	121
甘青铁线莲	33
甘肃马先蒿	130
甘肃丹参	121
甘肃棘豆	77
甘草	69
甘遂	223
艾蒿	168
节节草	2
本氏针茅	198
本氏蓼	10
石刁柏	209
石龙芮	221
石龙胆	106
石竹	28
石防风	104
石沙参	141
石菖蒲	203
龙须菜	209
龙胆	106
龙葵	127
平车前	137
打碗花	110
东方铁线莲	32
东方蓼	6
东北川续断	214
北水苦荬	130
田边菊	153
田旋花	110

生地	133
禾萱草	174
仙人果	47
仙鹤草	48
白三叶	57
白山菊	154
白车轴草	57
白头翁	44
白皮锦鸡儿	75
白羊草	184
白花苜蓿	54
白花刺	81
白花草木樨	58
白花蓼	8
白茎盐生草	15
白茅	190
白刺	88
白屈菜	222
白草	184
白茨	88
白背	154
白绒蓼	7
白莎草	202
白麻	95
白颖苔草	200
冬青叶兔唇花	121
冬青兔唇花	121
冬巢菜	61
冬葵	97
冬箭筈豌豆	61
鸟足豆	59
鸟蓼	7
兰花花	116
半扭卷马先蒿	131
半夏	225
半灌木	20
尼泊尔蓼	9
对叶草	69
母猪刺	75
丝裂亚菊	163

六　画

吉氏蒿	171

老芒麦	173	早熟禾(瓦巴斯)	181
老鸦翎	44	先锋高丹草	199
地丁草	213	竹节草	7
地八豆	225	血生	123
地构叶	92	向阳花	149
地构菜	92	合头草	13
地肤	17	合欢	81
地黄	133	伞花雅葱	160
地梢瓜	109	杂交(甜)高粱	200
地椒	120	杂配藜	24
地槐	217	多节雀麦	175
地榆	49	多年生黑麦草	180
地锦	91	多茎委陵菜	46
地蓼	7	多枝柽柳	92
芨芨草	186	多刺绿绒蒿	36
芒草	196	多变小冠花	59
西山委陵菜	47	问荆	1
西升麻	34	羊角菜	167
西来稗	191	羊茅	180
西伯利亚远志	90	羊蹄	11
西伯利亚蓼	8	羊蹄叶	11
西黏谷	26	并头黄芩	124
百合	209	米口袋	70
百里香	120	米瓦罐	30
百脉根	59	兴安毛莲菜	152
百蕊草	215	兴安虫实	22
灰叫驴	155	兴安胡枝子	72
灰灰菜	23	冰草	175
灰条	23	异叶败酱	139
灰菜	23	羽裂莛子藨	204
灰绿黄堇	222	羽叶点地梅	105
灰绿碱蓬	12	红三叶	56
灰绿藜	17	红山草	198
灰蒿	171	红车轴草	56
灰碱柴	20	红毛公	196
达乌里胡枝子	72	红丝草	91
列当	135	红刚芦	196
扫帚苗	67	红花角蒿	134
扫帚菜	17	红花岩黄芪	82
尖头叶藜	16	红花酢浆草	85
光药大黄花	134	红豆草	56
光雀麦	174	红果龙葵	127
早开堇菜	98	红狐茅	177

红虱	93
红柳	92
红砂	93
红柴	196
红蓼	6

七　画

麦瓶草	30
苇	197
苇状羊茅	177
苇状狐茅	177
苇状看麦娘	193
苋菜	26
花花柴	153
花苜蓿	73, 74
花柴	63
花棒	63
花帽	63
花蓼	8
苍耳	157
芦	197
芦苇	197
芦笋	209
苏丹草	199
苏枸杞	126
苏联猪毛菜	21
杠板归	11
杠柳	109
杏仁菜	91
豆寄生	113
两歧飘拂草	203
两栖蓼	5
旱苗蓼	6
旱型两栖蓼	5
旱莲草	148
园草芦	176
串叶松香草	142
串地龙	214
串地蒿	172
串珠芥	41
秃疮花	221
佛豆	65
谷莠子	196

角果碱蓬	22
角茴香	35
迎春花	216
忘忧草	205
闵草	192
冷蒿	172
沙打旺	57
沙米	18
沙枣	94
沙茴香	101
沙葱	206
沙棘	93
沙蓬	18
沙蒿	170
阿氏旋花	111
阿尔泰狗娃花	152
阿尔泰紫菀	152
阿拉伯婆婆纳	133
陇东苜蓿	53
附地菜	114
忍冬	204
鸡儿肠	153
鸡肠草	114
鸡冠花	97
鸡峰山黄芪	66
鸡脚草	178
驴食豆	56

八　画

武竹	208
青杞	128
青海黄芪	83
苦豆子	78
苦马豆	78
苦瓜	95
苦豆根	217
苦刺	81
苦参	217
苦荬菜	156
苦菜	156
苦蒿	155
苦房草	196
苘麻	95

直立黄芪	57		狐尾槐	78
茎草	184		狗牙根	189
枇杷柴	93		狗舌草	167
松叶草	139		狗尾草	195
枪草	183		饲料酸模	12
刺儿菜	146		夜合树	81
刺沙蓬	21		卷丹	205
刺蓬	21		油蒿	170
刺藜	16		沿篱豆	64
拐轴鸦葱	160		泡泡刺	88
顶羽菊	155		泥胡菜	149
抱茎苦荬菜	151		泽漆	223
拉拉藤	4		降龙草	4
拌根草	188		细叶亚菊	164
拧条	76		细叶百合	210
披针叶黄花	74		细叶远志	90
轮叶马先蒿	131		细叶刺针草	144
歧序唐松草	31		细叶益母草	119
虎尾草	184		细枝岩黄芪	63
果园草	178		细绿萍	2
败酱草	140		细裂叶莲蒿	169
垂序商陆	215		绊根草	189
垂果南芥	42			
垂果蒜芥	41		**九　画**	
垂穗披碱草	174			
牧场黑麦草	180		春黄芪	66
委陵菜	44		春巢菜	60
爬地草	189		珍珠梅	50
金色补血草	105		革命菜	143
金色狗尾草	194		茜草	138
金灯藤	113		草木樨状黄芪	67
金花草	58		草玉梅	219
金佛草	161		草本槐	217
金梅	216		草芦	176
金匙叶草	105		草麻黄	3
金银花	204		茴茴蒜	218
金银藤	204		茵陈	169
金腰带	216		荠荠菜	36
金露梅	52		荠菜	36
念念	87		茭蒿	171
念珠	41		荒漠锦鸡儿	75
肥马草	183		荨麻	5
兔毛蒿	172		胡豆	65
			药王茶	52

药瓜皮	95
枸杞	125
柳叶沙参	141
柳叶刺蓼	10
柳叶菜	100
柳枝稷	186
柳穿鱼	128
歪头菜	69
砂引草	109
挂金灯酸浆	126
星星草	193
香附子	201
香青兰	123
香铃草	96
香蓼	10
香薷	117
秋鼠麴草	159
鬼子姜	162
鬼针草	143
待宵草	100
胖姑娘	153
匍茎通泉草	129
匍枝委陵菜	45
狭叶米口袋	70
独行菜	39
弯曲碎米荠	37
美丽胡枝子	82
美国香豌豆	61
美国籽粒苋	27
美国商陆	215
籽粒苋	25
洛氏锦鸡儿	75
洋麦	181
洋姜	162
穿龙薯蓣	214
窃衣	103
扁竹	7
扁竹花	211
扁豆	64
扁苜蓿	74
扁核木	51
扁蓄豆	73
费菜	43

绒毛绣线菊	49
绒花树	81
骆驼蓬	89
骆驼蒿	89

十　画

秦艽	107
秦岭沙参	142
秦岭野韭	207
珠芽蓼	9
蚕豆	65
栽培山黧豆	62
盐爪爪	20
盐生草	15
耆状亚菊	172
莲山黄芪	65
荻	196
恶实	162
桂竹糖芥	42
桔梗	140
栝楼	95
夏至草	118
柴胡	102
鸭茅	178
鸭跖草	3
蚓果芥	41
圆叶牵牛	112
圆叶锦葵	96
圆形苜蓿	55
圆盘荚苜蓿	55
铁丝草	176
铁扫帚	72
铁苋菜	91
铁线草	189
铃铛草	179
积机草	186
笔头草	1
笔管草	160
倒提壶	115
臭草	183
射干鸢尾	211
狼牙刺	81
狼耳朵	133

狼尾巴花	104	菊苣	147
狼尾珍珠菜	104	萱草	198
狼尾草	194	梅花草	213
狼毒	224	梯牧草	178
狼紫草	115	梭梭柴	24
皱叶酸模	11	掐不齐	66
皱果苋	27	救荒野豌豆	60
高山早熟禾	182	雀野豆	80
高羊茅	177	野大豆	68
高冰草	185	野大烟	35
高秆菠菜	12	野芝麻	122
席萁草	186	野西瓜苗	96
离子芥	37	野苋	27
唐古特白刺	88	野苋菜	26
唐本草	144	野苏子	122
益母草	119	野苜蓿	54,58,73,74
宽叶香蒲	173	野茄子	128
宽叶独行菜	40	野茼蒿	143
家稗	187	野荞麦	9
展枝唐松草	31	野胡萝卜	101
通泉草	129	野胡麻	212
绢毛胡枝子	72	野韭	207
		野高粱	199
		野菊	155
		野麻	108

十一画

黄毛棘豆	217	野绿豆	68
黄瓜香	49	野葵	97
黄花矾松	105	野棉花	33
黄花角蒿	135	野罂粟	35
黄花苜蓿	54	野豌豆	80
黄花草木樨	58	野稷	187
黄花菜	205	野燕麦	179
黄花酢浆草	85	野糜子	187
黄花蒿	168	悬钩子	50
黄芩	125	曼陀罗	225
黄金茶	125	蛇莓	44
黄帚橐吾	166	蛇倒退	11
黄背草	198	铜锤草	85
黄蒿	169	银灰旋花	111
黄鹌菜	158	银露梅	52
黄缨菊	165	甜高粱	200
萝卜艽	107	犁头刺	11
菊叶香藜	14	犁头草	99
菊芋	162		

假苇拂子茅	197
假绿豆	74
盘棋	184
脱毛天剑	111
猪毛菜	21
猪毛蒿	169
猪殃殃	138
猫儿眼	223
猫耳刺	75
猫尾草	178
麻叶荨麻	5
麻花头	161
旋覆花	161
旋鳞莎草	202
剪刀股	8
牻牛儿苗	86
宿根亚麻	212
宿根黑麦草	180
密花香薷	117
蛋白草	27
绳虫实	13
绵团铁线莲	34
绵毛酸模叶蓼	7
绵蓬	22
绥草	212
绿升麻	34
绿苋	27
绿珠藜	16
绿穗苋	27

十二画

琥珀千里光	164
款冬	157
散生木贼	1
葎草	4
葡根骆驼蓬	89
落叶灌木	216
萹蓄	7
朝天委陵菜	45
葭	197
棒槌草	184
棋盘花	97
棉条	63

酢浆草	84
硬毛果	80
硬阿魏	101
硬质早熟禾	176
裂叶马兰	153
紫云英	60
紫羊茅	177
紫花山莴苣	151
紫花地丁	99
紫花苜蓿	53
紫花槐	63
紫苏	120
紫苜蓿	53
紫草	116
紫萱	205
紫蝴蝶	211
紫穗槐	63
遏蓝菜	39
景天三七	43
黑老鸦头	214
黑麦	181
黑沙蒿	170
黑果枸杞	126
鹅绒委陵菜	47
鹅绒藤	108
鲁梅克斯	12
湖南稷子	187
登相子	18

十三画

鹊豆	64
蓝花棘豆	83
蓝草	181
蓝蝴蝶	211
蓬子菜	139
蒿	171
蒺藜	87
蒺藜状苜蓿	55
蒺藜梗	18
蒲公英	156
蒙山莴苣	151
蒙古葱	206
蒙古雅葱	167

蒙古蒿	171	蝎虎草	87
蒙葱	206	蝎虎霸王	87
赖草	183	墨草	148
雾冰藜	14	稷子	187
蛾眉豆	64	箭叶橐吾	166
蜀葵	97	箭筈豌豆	60
鼠掌老鹳草	86	熟地草	188
微孔草	116	鹤虱	114
腺毛唐松草	219	缬草	140
腺毛翠雀	220		
腺梗豨莶	144		

十六画

燕麦	179
薄荷	124
糙叶黄芪	66
糙苏	122
糖高粱	200
藓状马先蒿	132

十四画

截叶铁扫帚	72
聚头蓟	165
聚合草	113
蓼子朴	150
酸枣	43
酸味草	85
酸胖	88
酸模叶蓼	6
碱地肤	18
碱韭	206
碱蓬	12
蝇子草	31
鲜卑花	51
獐牙菜	107
辣子草	149
漏芦	163
翠雀	220

十七画

藁本	103
蟋蟀草	189
黏毛蓼	10
繁穗苋	26
藜	23

十八画以上

藤长苗	111
蔍草	201
瞿麦	29
镰叶鸢尾	211
镰荚苜蓿	54
鼬瓣花	122
翻白菜	44
翻白藜	17
鳍蓟菊	154
瓣蕊唐松草	32
鳞叶龙胆	106
鳢肠	148
蘋草	176

十五画

耧斗菜	218
鞑靼滨藜	20
蕨麻	47
豌豆	64
醉马草	226
播娘蒿	38
蝎虎驼蹄瓣	87

参考资料

[1] 李扬汉.植物学.上海:上海科学技术出版社,1984.
[2] 任继周.草业大辞典.北京:中国农业出版社,2008.
[3] 富象乾等.植物分类学.北京:中国农业出版社,2003.
[4] 贾慎修等.中国饲用植物志.北京:中国农业出版社,1987.
[5] 侯宽昭.中国种子植物科属词典.北京:科学出版社,1998.
[6] 高信曾.植物学.北京:人民教育出版社,1978.
[7] 王枝荣.中国农田杂草原色图谱.北京:中国农业出版社,1990.
[8] 闵乐园.陕西牧草.西安:西北大学出版社,1987.
[9] 付坤俊.黄土高原植物志.北京:科学技术文献出版社,1989.
[10] 马奇祥等.农田杂草识别与防除原色图谱.北京:金盾出版社,2005.
[11] 刘媖心.中国沙漠植物志.北京:科学出版社,1985.
[12] 罗献瑞.实用中草药彩色图集.北京:世界图书出版社,1998.
[13] 邢旗等.内蒙古草原常见植物图鉴.呼和浩特:内蒙古人民出版社,2008.
[14] 贠旭疆等.中国主要优良栽培草种图鉴.北京:中国农业出版社,2008.